The Kindness of Human Milk

The heart of human health

Larry Churchman

This book is copyright © Larry Churchman 2017 and must not be copied, reproduced, transferred, distributed, leased, licensed, publicly performed, or used in any way except as specifically permitted in writing by the publisher, as allowed under the terms and conditions under which it was purchased or as strictly permitted by applicable copyright law. Any unauthorised distribution or use of this text may be a direct infringement of the author's and the publisher's rights.

ISBN: 978-0-244-94113-0

Published by AtopicalPublishing.com
Cambridge UK
2017

Acknowledgements

Numerous friends, colleagues and family members have offered advice and encouragement over many years during the preparation of this book. I feel especially indebted to my wife and daughters, who have indulged my single-minded pursuit to share what I have learned from dealing with my own atopic disease symptoms and those of friends and colleagues.

This book has an associated website:
https://thekindnessofhumanmilk.com/
Join the conversation and find related material.

Contents

Introduction	1
The largest unrecorded animal experiment in history	3
Atopic symptom aetiologies	26
Diet-related psychotropic effects	59
Social consequences	66
Research into diet-related disease outlined	73
Resolution	85
Summary statements:	97
Glossary	98
Bibliography - Books	102
Bibliography - Papers	103

Introduction

NB This book is purely informational. It should not be interpreted as a guide to self-treatment of any kind. All readers are recommended to rely on the advice of a qualified practitioner for any medical or psychological ailment. The author and the publishers accept no legal liability for outcomes resulting from the interpretation of the content as constituting medical guidance.

What has caused Western illnesses? Why do they appear to be becoming more widespread? What might be done to reverse this trend?

Much has been written around this subject, but there have been few convincing attempts to integrate observations with explanations. In this book I attempt to explain how atopic conditions such as asthma, eczema, autism, and many others, can be traced back to the failure to exclusively breastfeed children for the first months of their lives.

I have spent around thirty years wondering and investigating whether or not such changes in the human organism, which manifest as Western diseases, can truly be explained by the advent of feeding newborn babies on confected formula milk. The idea that just one change in human nutrition could spawn such a range of diseases seems so improbable, but observational, research, and treatment findings indicate a strong likelihood that this is the case.

Like so many people today, I have suffered a range of atopic symptoms from an early age. Migraines had begun to affect me by the age of two, and I have strong and detailed recollections of at least one such event at this age. My bursting appendix had to be removed in a hurry at the age of nine. This was a consequence of coeliac disease, which was beginning to make my life uncomfortable, but which had not yet been specifically diagnosed.

Nasal and bronchial congestion became a feature of the next ten years, along with eczema. Sinus problems ensued. At around the age of thirty-four, the first joint pains signalled the start of rheumatoid arthritis, which eventually took me out of a career in teaching.

Connecting diet with arthritis helped me towards freedom from those symptoms. The foods that caused my rheumatoid arthritis were identified and the disease could be switched on or off at will, by eating or avoiding the implicated foods. Just prior to this revelation, I had been discussing arthritis options with my local GP and had rejected the offer of various pharmaceuticals that 'might be useful' but were not curative, and which may have caused considerable harm if taken over a prolonged period.

Understanding the origin of atopic disease is a starting point for reducing its impact.

I believe that the total worldwide damage to quality of life, and the pain and the early deaths inflicted by the use of manufactured milk formula to feed newborn babies, overwhelmingly exceeds that of the impact of the nuclear weapons exploded over the defenceless populations of Hiroshima and Nagasaki.

This book attempts to challenge many of the widely accepted medical and psychological paradigms associated with Western illnesses. It provides a different way of thinking about modern somatic and psychological illness, based on common observations by families, medical practitioners, and researchers over the last two centuries. This is not a scientific document, but a short book of ideas which have arisen from my experience of atopic illness and its effects over seventy years of my life.

The largest unrecorded animal experiment in history

The essence of human milk

In many different manuals on childbearing, whether from the orthodox medical approach or from the approach of alternative medical thinking, there are confused statements about the benefits and the disadvantages of human milk for the health of newborn infants.

The arguments presented here are based on the immunological processes that occur between the mother and the child and between the child and its environment. Many of the ideas put forward here are based on original thinking, and are either new or presented in a new light.

The author believes that human milk is best, but for rather different reasons from those normally put forward. Maternal milk is good for a newborn infant. This is not because of any specific ingredient and not for any psychological impact, but it is simply because it does not provoke an immune reaction in the way that all other foodstuffs will.

At birth, the infant immune system generally continues to treat maternal antigens as self. During gestation, it is important that the foetus does not begin an immune reaction against the mother and that the mother's immune system does not attack the foetus: their immune systems adjust accordingly. Immunologically, mother's milk will normally be treated as self by the infant, and will provoke no immune reaction.

Any non-maternal milk, including formula milk, will be treated as a potential pathogen by the immune system of the immature gut. It will provoke a reaction, which, in time, will result in the production of antibodies against some of those food components. This will then start various atopic disease processes. The very assimilability of the components of cow's milk in formula feeds ensures that the antigenic components reach deep into the newborn child's body systems. The immature gut does more assimilation than digestion, and so the first few meals pass largely unchanged into the infant's circulation. This is good if the milk comes directly from the child's mother, but not if the milk is from another, unrelated species of mammal or plant. The same is true for the newborn of most mammal species. Most vets know this important fact.

Much of the influence of medical orthodoxy is negative in terms of natural processes, and usurps parental involvement. Medical fashions tend to be relatively short-lived and parents feel unable to employ the knowledge they have picked up from their social environment, because it has become 'out of date'. Examples include the posture for giving birth, whether to eat or not eat liver when pregnant, and whether bottle-feeding is a neutral option or a more damaging option.

Medical orthodoxy and technology have combined to produce a greater negative impact on the natural processes of human reproduction and health in industrialised nations. Their influence has also spread much of this negative impact on the same processes to less developed nations, sometimes in the name of overseas aid and development. Reducing infant mortality is not so much a bonus when the survivors and successive generations face a life burdened by Western diseases.

There is a need to rediscover the more natural elements of human reproduction and infant development before this knowledge becomes entirely lost to our civilisation. Our civilisation is just beginning to understand the impact of atopic disease on our communities and economies and, especially, the burdensome impact on health services.

Loss of kinship

In surviving older cultures of humankind, where kinship plays an important part in many activities and decisions, there are solutions that serve the greater need of the community in a simple and successful way.

One facet of these cultures is where a child is born to a mother who is unable to feed her baby because she can produce little or no breast milk, perhaps due to illness or even death. Within the clan, there may be close family members who are able and willing to undertake the task of breastfeeding, and the child is then able to thrive on affection and human milk. Without such intervention, in the absence of alternatives, the child would probably die.

Western societies began to change all this, partly because of prudery and partly through the application of the developing sciences. In history, from at least 2000 BC there has been the tradition of hiring a wet nurse to feed an infant, but the choice to pursue this action was often based upon criteria of class and financial status, rather than need.

Scientists first became involved in decisions on infant feeding based on their 'scientific' analysis of human milk. Their strange conclusion, at the time, was that human milk was insufficiently nutritious to support the healthy growth of an infant. We now know that the key milk samples may have been taken in ignorance of the dual nature of milk flow. A first thin, sweet milk flow is followed by the let-down reaction in which creamier, richer milk is produced after a period of comfortable feeding. Unfortunately, 'science' had 'proved' that there may be faults in human milk and the action was on to find 'scientific' substitutes.

The assumptions that were made by these scientists about the survival of humans to that point in time are beyond comprehension, but the expression 'blinded by science' comes to mind. Thus began the history of 'scientific formula milk' products such as Liebig's formula, which was marketed at first as a liquid consisting of cow's milk, wheat flour, malt flour, and potassium bicarbonate, and subsequently sold in dry powder form, to be mixed with diluted cow's milk.

Nutritional science was developed relatively recently in the West, to help plan the feeding of large populations during major wars and afterwards. The influence of continuing nutritional science based on biochemistry and biometrics has been revealed more recently by the World Health Organization. This organisation identified that its own pre-2006 percentile tables for infant growth were largely based on bottle-fed infants and had possibly led to widespread obesity in children, who had been overfed to make them conform to the percentile tables. Because such growth rates cannot easily be supported by human milk, the tables also exerted a pressure to bottle-feed. The author believes that bottle-feeding is the major cause of obesity in children and adults – not through a simple nutritional imbalance, but through more complex immunological processes.

It is human to experiment, except when bound by custom. It is therefore likely that over the last million years the feeding of infants has followed a somewhat varied pattern in different human cultural groups around the world. Nothing is likely to have been on the same scale as the present consumption of formula milk, but records over the last 4,000 years indicate that there were several cultures in Europe that introduced various ways to contaminate, or waste, either the first or many of the natural neonate feeds.

It was explained to me by a Sikh family living in the UK that they had tried to recreate their traditional customs after arriving in Britain following a period in Africa, where they had become very Western in their ways. One of the customs that they had held dear was the feeding of a child with samples of adult food at the time of the 'first grains' ceremony. In their recreation of this ceremony, they had slipped the timescale back towards the immediate postnatal period, though traditionally the first grains are fed when infants reach about four to six months. In this family, very young infants were ceremonially fed samples of adult foods with a silver spoon. These foods included yoghourt and cereals. When I first met the family, there was hardly a child that was not either eczematous, or extremely hyperactive, or both. There were young adults with symptoms of asthma, arthritis, or hypertensive disorder. The few members of this family in relatively good health were the elderly, who had been born and raised traditionally in India.

To me, this illustrates a change in infant diet based upon ideas of cultural identity and religion, rather than on any practical or dietetic reasoning. This has happened many times within urban societies. Such changes are unlikely to become widespread within any distant rural community because the impact on health would reduce a family's viability in such an environment. The bloodline would be likely to die out.

Expansion

At first, the use of formula milk was restricted to those who saw it as a more easily managed alternative to wet-nursing and to those who came under medical care, at or just after birth, because of circumstances such as the death or extreme illness of the mother. In time, the use of formula milk became more widespread as a result of the influence of the public advertising of different brands. Today the echoes of that advertising can be heard in the advice of some paediatricians: 'At least we know what is in formula milk.' This is in comparison to breast milk, which is perceived to be nutritionally variable and mysterious. Indeed, breast milk is variable in composition throughout an infant's early life, as it adapts to the child's nutritional requirements. When there are difficulties with breastfeeding, all that may be required is to show the mother how to help the child suckle effectively to promote effective lactation.

There is an unfortunate consequence of giving formula feed to newborn infants. It reduces the ability of that new generation to breastfeed. One reason for this is that bottle-feeding tends to result in various atopic skin problems in childhood and adulthood. These skin problems can range from dryness of the skin to full-blown eczema or psoriasis. This tends to make nipples more prone to damage after periods of breastfeeding, and produces discomfort and cracked and bleeding nipples. There may also be insidious changes to the internal, mucosal tissues of the breast. This damage is often sufficient to put the mother off breastfeeding entirely, with further consequences for the next generation. The result may appear to epidemiologists to be a simple, cultural trend. Today, in the UK, some young women see breastfeeding as disgusting and are determined to avoid doing it themselves, though this may simply be an attitude of youth bolstered by feminist perspectives.

At present, very few infants go through to weaning age without having tasted formula milk of one kind or another. Formula milks have changed over time in response to new nutritional science theories, or suspicion that previous formulations were responsible for some health issues (for example, the short phase during which fairly crude peanut oil was added as a lipid replacement component to one brand of formula milk, before the product was hastily withdrawn from the market).

Many infants fed on Liebig's formula, the first commercially available 'milk' formula, died after a short and miserable life. Unfortunately, it was thought that finding an improved formulation was the way to progress. There could be no consideration of the impact of formula feeding on the immune system or the later life of the individual, nor any thought that the correct next step was to pursue the kinship option of our ancestors. With subsequent reformulations, infant deaths continued, but were less frequent than with Liebig's original commercial formula. Sadly, the safety of formulations has been assessed only in terms of the effects on infant health. The potential impact on health in teenage and adult life has been largely ignored. The connection between infant nutrition and adult ill health has been a woefully neglected area of study.

Records

In all of this, there is one feature that should surely astonish all who depend upon science and medicine for assurance about health issues and infant care. Few medical practitioners or medical establishments have kept records relating to infant feeding over the last eighty years and more, except where special feeds have been administered in response to infants suffering ill health.

To put it another way, humans have become the experimental animal in the largest-scale nutritional experiment ever conducted, but for which few rigorous observations have ever been recorded, archived, or published.

While this experiment has proceeded – and it is far from finished – there has been an almost exponential rise in the burden of atopic disease on the populations affected: hence, the adoption of the expression 'Western diseases'. Now we are in a position where we cannot be sure of the extent to which this huge nutritional experiment has affected the population. We have little formally recorded evidence of an identified relationship (or absence of relationship) between the use of formula milks and atopic illness, for example. However, during this time, incidences of asthma, eczema, hyperactivity, glue ear, autism, obesity, eating disorders, early-onset diabetes, late-onset diabetes, irritable bowel, depression, arterial disease, rheumatoid arthritis, Alzheimer's disease, and some cancers, have steadily increased. The puzzle is to distinguish which of these have, or have not, been caused by the use of formula feed for newborn babies.

My personal experience directly links five of the above diseases to having received one formula feed within hours of my birth. This feed was given to allow my mother additional rest while sedated, in the hope that it would lead to more successful breastfeeding. These actions were part of an eighteen-month medically approved study carried out in the London-based nursing home where I was born. This intervention represents another short period of experimentation for which few records were kept, other than those that would determine eventual success with breastfeeding. I became aware of this experiment at about twenty-four years of age, through a chance meeting with a nurse who was working at the nursing home at which I was born and at the time that I was born. This person informed me about the timing of the experiment.

Despite this first feed of infant formula, my mother assured me that I was solely breastfed from birth to weaning some months later, as had been the case for my elder sister, who is still healthy to this day. My mother was unaware of the experiment carried out on herself and her son.

I can no longer eat the food groups represented in that formula feed without somewhat disabling symptoms, including rheumatoid arthritis and migraine. Avoiding those food groups, however, has led me to relatively good health, and I am still free of long-term prescription drugs at seventy years of age.

Experimental models

In 1978, Professor R. R. A. Coombs and his colleagues carried out an experimental series with guinea pigs, in which cow's milk was fed to neonate cavies from the second day of life. The reactions to this diet varied from apparent tolerance to the development of a potential for anaphylactic shock reaction to milk proteins, especially casein.

We cannot expect people to respond in ways exactly similar to guinea pigs, but this does demonstrate that cross-feeding of neonate mammals is not without risks to the integrity of the immune system. It also identifies the variability of response of siblings to a similar cross-feeding experience.

Some recent studies have identified a link between early-onset diabetes and the feeding of formula milk in certain North American populations. Identification of such links is now very difficult and runs contrary to our culture, in which toy dolls come with toy feeding bottles. It is not that there is no evidence, but that there are few remote populations of mankind remaining that are yet to be affected by the worldwide marketing of formula milk, or by medical intervention. There are very few control populations available for comparative studies, and those that do exist are likely to be too remote, too unwilling, and too inaccessible for controlled study.

Such populations as exist are likely to persist when the rest of civilisation is at risk from the weakening effects of atopic disorders. We may need to place our trust in their knowledge and customs for the future of mankind.

Inheritance and environment

We understand that two basic contributing factors make us what we are. They are our inherited constitution and the further shaping of that constitution by the environment in which we live. Environmental influences include, diet, culture, disease, parasites, stress, altitude, pollution, etc.

For a long time, the Japanese were held as an example population, with low rates of atopic disease and lower bowel cancer. Much was attributed to their fish-rich diet. In truth, the strong adherence to traditional infant nutrition and the lack of a dairy industry may have been the true reasons. As markets have opened and values that are more Western have been adopted, so Japan's incidence of atopic illnesses has risen sharply. Organised commercial dairy farming began during the Meiji Restoration (circa 1868), as Japan rebuilt its political and social structures to compete with the industrialised Western powers. Today Japan imports dairy products, especially from New Zealand and Australia, to supplement home production.

In Japan, exclusive breastfeeding at one month is now below 50 per cent of mothers and Western illnesses in Japan are on a steep rise, with lower bowel cancer becoming more common. Mental illness has also increased, with previously rare syndromes appearing across the population. The health authorities in Japan were alarmed at the rapid rise in asthma and pointed to various possible causes such as the rise in air pollution from traffic, which appeared to be correlated. The irony is that Western foods were incorporated into the Japanese diet to improve the general health of the population. Initially this did help, and the consumption of greater amounts of saturated animal fat helped reduce the incidence of cardiac failure in young Japanese men.

Inheritance through the maternal line

What tends to be overlooked is that we inherit features both genetically and non-genetically. The non-genetic inheritance is mostly through the maternal immune system, and involves the transplacental passing of antibodies and immunocytes to the near-term foetus. This provides it with immune programming and antibodies to face the infections that have been recently encountered by the mother-to-be. It may also help to prepare the gut mucosa to deal with the wide range of foods to be eaten in due course. This non-genetic aspect of inheritance can have a dramatic impact on our lives – for example, surviving a disease epidemic, which might otherwise have resulted in early death. There are other consequences, and not all of them are beneficial.

Today's fashionable scientific trend is the spectacular advance of genetics, including biomolecular studies linking genetic differences to physiological variants. The rapid advances in this area of science have allowed researchers to fall into the trap of believing that it is appropriate to solve physiological puzzles through genetic research. Commonly, the identification of a gene is taken as the identification of the cause of a physiological disease.

Life is rarely that simple. Genes are part of the mechanism for determining and maintaining an individual life form, but are not the full mechanism. Genetic research helps to illuminate how genes operate, but it does not replace an observational study of the whole organism.

The latest genetics research areas have included epigenetics (which concerns the cellular control of gene expression by epigenes) for many aspects of cellular activity. In the general sense of the term, atopic disorders are indeed epigenetic variations. In the stricter sense of the term, with epigenes controlling gene expression, it currently seems unlikely that asthma, eczema, and other atopic manifestations strictly fit the pattern. Rather, it seems that transferred maternal antibodies and maternal immunocytes affect the immune expression of the foetus and the neonate.

It may be the direct transplacental migration of antibody-producing cells, or cells carrying the specific antibody templates, that help to programme the infant immune system as part of a normal maternal microchimeric process. Where the maternal blood also carries the respective antigens that provoke the production of antibodies in the mother, some of these antigens may cross the placenta and become antibody targets for the foetal immune system to learn from. This 'learning' may involve the mechanisms of the epigenetic control of immune cell DNA.

Maternal inheritance has similarities and differences when compared with inheritance from the paternal line. The inheritance of acquired characteristics through the passage of immune components and the programming of the immune system from mother to offspring are the most significant. From the foetal perspective it may be viewed as an environmental inheritance.

The diagram below illustrates the simple inheritance of atopy through the maternal line. The inheritance is of the production of antibodies to commonly eaten foods, usually including foods of bovine origin. The resultant symptoms will depend upon the nutrition of the infant and the genetic input from the mother and father of the child.

Family tree showing a typical example of acquired atopic disease inherited through the maternal line by transplacental passage of antibodies in the absence of further bottle-feeding:

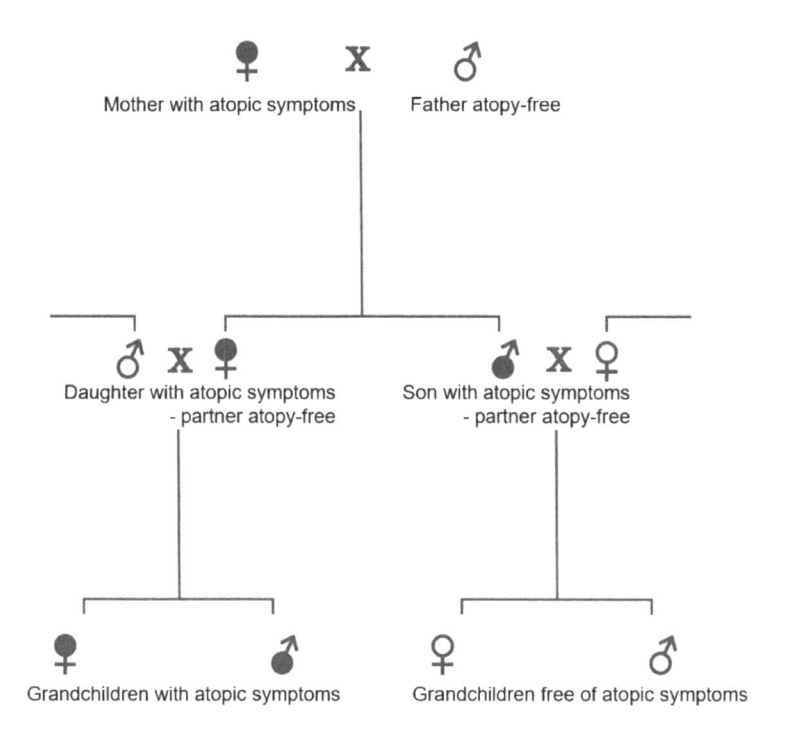

♀ X ♂
Mother with atopic symptoms ⸳ Father atopy-free

♂ X ♀   ♂ X ♀
Daughter with atopic symptoms   Son with atopic symptoms
- partner atopy-free   - partner atopy-free

♀ ♂   ♀ ♂
Grandchildren with atopic symptoms   Grandchildren free of atopic symptoms

Work has also been carried out recently on the contribution to health made by the genetic action of maternal mitochondrial DNA. This research may have been undertaken to explain the observation of the passage of atopic symptoms through the maternal line, but the most direct maternal contribution to the immune system of the subjects studied seems to have been ignored. Perhaps this is an oversight. The transplacental transfer of antibodies and immunocytes is the obvious mechanism for the passage of atopy from mother to child. This transfer has been called 'passive transfer of immunity'. However, it has recently been recognised that it is a far from passive process, and that there can be a degree of programming of the immune system of the recipient resulting from such a transfer.

The inheritance of immune characteristics that have been acquired through transplacental passage can affect subsequent generations by the same process, through the maternal line. Anti-food antibodies that are passed on to a child are likely to be reinforced by the reactions they induce. A female child may then go on to become the donor of such antibodies to her own offspring. And so, through the generations, each generation will produce the potential for a wider range of food intolerances. This is reflected in the rapid spread of atopic disease throughout populations in industrialised countries. The process is enhanced by the likelihood that diet and cooking choices are often passed from mother to daughter. Eating and cooking in the same way as one's mother, and in the same way as her mother before her, ensures that the environment for atopic disease is maintained.

Here we are examining a mechanism of inheritance that should promote the health of the individual by augmenting immune responsiveness. Indeed, it is very likely that those with the most vigorous immune response to infections, who should be among the healthiest in our population, are the ones to suffer most after acquiring antibodies to foodstuffs. They may go on to suffer more extreme symptoms. A strong complement reaction focused on infectious disease organisms will quickly clear an infection, but a strong complement reaction to antigen-antibody complexes, as a consequence of ingesting foodstuffs, is likely to provoke the most debilitating atopic symptoms.

I see a danger here that genetic studies will identify the genes associated with severe atopic symptoms and mistakenly describe them as the cause of atopy, instead of as the genes that would normally provide their owner with excellent immune competence, were it not for the nurturing catastrophe of feeding neonates with proprietary infant formula.

Darwin and the inheritance of acquired characteristics

It is interesting to reflect on Charles Darwin's own bouts of illness. As some of his biographers have suggested, were they psychosomatic symptoms? Or were they inherited atopic somatopsychic symptoms that were related to his mother Susannah's illness, which eventually proved fatal? Susannah appeared to suffer from an acute enteric illness, similar perhaps to Crohn's disease, though it was possibly cancer. There are indications that she may have been 'invalided' for an undefined period before the severe pain she experienced required opiates to reduce her suffering. Susannah died aged fifty-two in July 1817 from the effects of peritonitis, following 'perforation of the peritoneum'.

Did Darwin therefore suffer from symptoms of an illness that was inherited as an acquired characteristic through his mother Susannah's contribution to his foetal immune system, by the transplacental migration of her errant immunocytes and antibodies during gestation? Some have suggested that his illness was most likely to have been Crohn's disease. This might help to explain Darwin's later obsessive–compulsive behaviour in relation to his family, and especially his later intense need to perpetually research all the areas of science that had a bearing on his hypothesised evolutionary drivers. And it may also explain his decision to have the road adjacent to Down House lowered to stop local people staring into his garden and home.

These are all indications of an obsessive–compulsive mind that may have been perturbed by food intolerances. Somatopsychic symptoms of food intolerance can include obsessive–compulsive behaviour and depression. His physical symptoms certainly point to atopy. By his own description, he had occasional eczematous phases, he suffered from continuous periods of fatigue, he had occasional palpitations, and he was frequently nauseous with acid reflux. Such a range of physical symptoms is most unlikely to be psychosomatic. Both the onset date and the range of symptoms clearly indicate the breakdown of an adaptive allergic response, where the exhausted adrenal glands cease to provide sufficient corticosteroids to prevent the expression of physical symptoms.

The possibility is that Susannah, and even Charles, may have been fed with dairy milk, or other foodstuffs, at too early an age. Given that Susannah died from gastrointestinal symptoms, it is possible that she was fed on unsuitable foods soon after her birth. Baron von Liebig was widely advertising his 'Soluble Food – the most perfect substitute for Mother's milk' in 1869, having published *Animal Chemistry*, his first book on animal nutrition, in 1842. It is therefore likely that various scientific infant feeding experiments were in progress well before the marketing of Liebig's formula, especially in a family such as Darwin's, which had a line of inquisitive practising doctors of medicine.

On careful consideration, the records suggest that Charles Darwin may have been affected by his mother's transferred antibodies. In a culture where Freud's postulates are generally accepted, the diagnosis of a psychosomatic cause is the easier option and conforms to our inclination to place blame on the patient. However, the observed connection between diet and mental health contradicts Freudian notions. Here, observation appears to support the presence of food intolerance, probably to dairy and beef. Darwin fell ill when he was at home in England, where he had access to his favourite items of food, and not while he was on his remarkable voyage to remote parts of the world. Supplies of cheese and salt beef would not last a full voyage, thus removing most foods of bovine origin from Darwin's diet within a few months. His illness persisted during his more settled time at Down House.

Drugs and knowledge

In the West we are gradually coming to realise that possessing self-knowledge and taking personal control of our illnesses can, in many cases, be more effective than powerful drugs. If you have a tendency to late-onset diabetes in your family, for example, it is probably better to control your calorie intake and take good and regular exercise than to resign yourself to depending entirely on drug or insulin control of the key symptoms later in life. Incipient cardiac disease can often be reversed by attention to diet and aerobic exercise.

Very few diseases have no alternative treatment options to manufactured pharmaceutical chemicals. We have accepted the amazing lifesaving properties of some manufactured pharmaceuticals such as antibiotics. And, somehow, we have been persuaded to extend our experience of these drugs to a belief in the superiority of all manufactured pharmaceuticals, including many that are subsequently withdrawn from sale after maiming or killing an 'excessive' number of recipients.

Due to the high cost of developing and registering a new drug, the pharmaceutical manufacturer is focused on the continual sale of that drug and not on ways of overcoming an underlying disease process. In fact, there have recently been authoritative assertions that pharmaceutical companies have invented illnesses, or implemented 'disease branding' to promote or boost sales of their existing registered drugs.

When we opt for modern manufactured drug treatments, we are effectively resigning from a sphere of life knowledge and allowing that knowledge to be usurped by the pharmaceutical industry. Even doctors become limited to the knowledge that the drug companies are prepared to offer them about current pharmaceuticals. The consequences of this are occasionally tragic, as were the consequences of the marketing of thalidomide as a sedative and an anti-emetic for pregnant women.

To regain control of our own treatment, especially with regard to atopic illness, it is important to reverse this trend and to rebuild a working knowledge base for personal and individual application. Present funding and regulatory models do not facilitate such an approach. In fact, they deliberately oppose it. This is a result of the lobbying power of the drug industry and the medical establishment in retaining control of orthodox medical knowledge and practice.

A new victim of this lobbying power will be the ancient and established practice of herbal medicine. It is from this ancient tradition that many pharmaceuticals have been developed – not necessarily to improve their efficacy, but to produce drugs that can be patented because of their unique synthetic chemical attributes.

Atopic illness accounts for a major part of health service expenditure, due to the long-lasting nature of atopic symptoms such as asthma. Drug companies have a strong commercial reason for being opposed to the unravelling of the atopic disease web. One of the dangers with atopic illness is that 'ethical pharmaceuticals' are now available to block some, or even most, of the symptoms of the disease, while allowing the underlying disease processes to continue (an example is the typical sort of asthma-relieving inhalant).

Some practitioners now accept that the early treatment of asthma by this means can make the disease more dangerous and that respiratory control exercises can bring about a better resolution in some younger sufferers, and also in sufferers over the longer term. Additionally, the use of inhalers can induce dependency. Nevertheless, asthma is just one symptom of a complex disease, and controlling asthma pharmaceutically does not usually halt the underlying disease processes. These disease processes are likely to eventually produce obesity, cardiac disease, mental illnesses, cancer of the gut, the breast, or the prostate, Alzheimer's, or other catastrophic symptoms.

The typical atopy pattern: evolution of symptoms

In the newborn child, eczema is often taken as the first sign of atopy, though it may be preceded by other atopic symptoms that are not seen to be associated with the eczema itself. Many of the indications of atopy are behavioural: unusual sleep patterns, erratic crying, or long periods of silent gazing.

When a baby is weaned from the breast it is often observed that he or she will vomit their first few intakes of formula feed. This vomiting is often quite violent and fits well the descriptive term 'projectile vomiting'. Additionally, there may be colic before or after the onset of bottle-feeding, accompanied by unusual sleep patterns. Some infants show eczematous skin at or soon after birth. Many show symptoms of eczema soon after weaning onto formula milk. These manifestations indicate the transfer of the mother's anti-bovine antibodies, transplacentally and through breast milk.

A hundred or so years ago, most of these incidences would have been rare, yet today they are considered normal. Old fears of variation from the norm were probably based on generations of observation that linked different variations of skin and alimentary conditions with crippling disease occurring later in life. There would be a distinct difference between osteoarthritis caused by life's wear and tear and rheumatoid arthritis. In one case, the body responds as might be expected to an excessive physical burden and poor diet, and in the other the body creates problems without the influence of a physical burden. The appearance of eczema early in life may have helped to foretell suffering later in life.

Pregnancy, birth, and childcare were once managed entirely by women, and within a family or a household. Families would have observed all the features of each child in detail and would have been able to continue that observation throughout much of the child's life, passing on observations to each other and from one generation to another. Under such a system, it becomes easy to associate future health with observations from pregnancy onwards.

Modern family structures and health systems work against such observation. The result has been a failure of knowledge integration. Assessments of the safety of changes to formula milk composition have been based on the effects on the infant, not on the potential effects over a whole lifetime.

Divisions within modern medicine have tended to leave the analysis of eczematous skin conditions to dermatologists, when perhaps the involvement of paediatricians, gastroenterologists, and rheumatologists would have been helpful. Many gastroenterologists come from a surgical background that denies them time to study the full immunological complexities of the function of the gut. Skin specialists tend to be absorbed in the pharmacology of variations of cortisone-based preparations or in the analysis of precancerous and cancerous skin conditions. Paediatricians are more commonly absorbed in the conformance of growth and developmental patterns to published data than in the immunological basis of childhood disease. It has therefore become difficult for any medical specialist to take a wider and more general view of what is happening to patients within such a system. The knowledge integration process hardly exists.

Patients who do not fit within the parameters set by the organisational matrix of medical practice do not fare well and may be passed on to psychiatric specialists, as if it must be the patient's own mental state that has caused them to fail to fit the medical treatment system. Such psychiatric interventions are rarely helpful and can result in harm to the patient, especially when psychiatric practitioners adopt the role of 'ethical pharmaceutical resellers'.

The atopy pattern starts with a child who may have suffered various symptoms of eczema and colic from before or immediately after weaning. The colic and eczema tend to continue for at least a year and new symptoms tend to replace those that fade. Glue ear, tonsillitis, and nasal congestion often come to prominence in the next phase, along with frequent bronchitic problems, possibly asthma, and some behavioural issues or learning difficulties. Thankfully, not every affected toddler displays all these characteristics and the order of symptoms may vary, but the outcome is similar. An atopic child becomes somewhat encumbered by the symptoms and the attention that is required to cope with them.

The insertion of a grommet into the eardrum, to relieve the distress of glue ear, became one of the most frequent surgical interventions for young children in the UK in 1986. Use of this surgical technique waned as alternative ways of managing glue ear developed. It has the merit that it protects against pressure within the ear, which could cause long-term deafness. However, the relief given by the operation tends to obscure the fact that there is still an underlying and ongoing disease process, and that only one symptom of that disease process has been relieved by the operation. Other symptoms such as fatigue and occasional sinus and chest infections are less defined and less easy to picture as part of a pattern, but they are usually present.

The prescription of antibiotics is often highest among those with atopy, and in time this may affect the microorganisms inhabiting the complex microbiome of the large intestine. The persistence of perceptional and/or behavioural problems tends to confuse the picture. The symptoms of disturbance may simply show as persistent unruliness but may also extend, for example, to very clear-cut autism.

As the teenage years approach, early medical intervention appears to have been rewarded with an apparent reduction in symptoms, or of the severity of symptoms. The grommets may be removed, the asthma is perceived as being under control by corticosteroids, the infections have been suppressed by using antibiotics, and the eczema has been suppressed, also by corticosteroids. However, unseen changes are taking place that will affect the child and the adult throughout the rest of his or her life.

The atopic disease processes are complex, but the causes appear to be quite simple. There appears to be a single switch that turns on atopic illness for life. If this switch is not turned on then the person will be left free of such disease. Some atopic disease sufferers appear to be marginally affected, though this may be a matter of perception rather than of reality. Other diseases related to those already described include early-onset diabetes, depression, rheumatic illness, psoriasis, osteoporosis, Crohn's disease, irritable bowel, hypertension, atheroma with atherosclerosis, tonsillitis, inflamed appendix, ankylosing spondylitis, etc. ... The range of symptoms and outcomes seems suspiciously wide. The onset of apparent disease symptoms often occurs in midlife, when the adrenals cease to continue to be able to produce sufficient corticosteroid to suppress atopic symptoms.

What switch could bring such a list of miseries to a life? How have we allowed such enduring pain and life-restricting horrors to become the norm for a significant and rising percentage of the population? Who should have been protecting the population from such an onslaught of misery? Who is ultimately responsible?

The origin and history of the bottle-feeding of infants is closely linked with science and medicine. Scientists determined the nutritional necessity of supplementary feeding, and medicine was instrumental in promulgating the fashion for bottle-feeding. Food scientists determined the formulation of various feeds, and nurses demonstrated the routines to be followed to expectant women and new mothers. Nobody within the system seems to have anticipated the extraordinary effect on Western populations and, subsequently, on most populations around the world.

Atopic symptom aetiologies

Asthma

Asthma is a complex disease that has significantly increased in terms of the percentage of the population affected by it. It appears to be quite obvious that exposure to certain substances in foodstuffs, or in the air, causes wheezing and breathing difficulties. Therefore, asthma is prevented by avoiding such irritants. However, the true pattern of the disease is more complex.

Asthma is a symptom of an underlying disease process and an asthma attack can be triggered by a range of irritants, but the cause of asthma lies much deeper. From my own observations, asthma is generally the result of an immune intolerance to bovine-sourced foodstuffs – in other words, products and by-products of milk and beef. We should not be surprised that milk intolerance can extend to beef intolerance, because these foods are parts of one species of animal.

An intolerance to bovine foodstuffs is initially created through the consumption by neonates of infant feeding formula based on cow's milk. This intolerance loads the bloodstream with milk-based antigen-antibody immune complexes. These complexes can respond to irritants through the generation of local histamines at the site of irritation. The immune complexes become locally adherent and bring about a sequence of reactions known as the complement series reactions. This amplifies the original impact of the irritation exponentially, resulting in gross respiratory distress.

One might think, therefore, that the answer to asthma is to stop ingesting foodstuffs of bovine origin. Indeed, that does work for many people. Even reducing the level of consumption or changing the type of food, for instance, condensed milk instead of pasteurised milk, can remove or reduce the symptoms of asthma. However, there is a danger in this course of action.

David, a former office colleague of mine, had suffered from asthma for many years, together with sinus congestion and other minor symptoms. Spring was approaching fast and he feared his asthma trigger – freshly cut grass. I advised him to reduce his intake of milk and beef-based foodstuffs. To his surprise, he found he could control the severity of his asthma symptoms by increasing or reducing his consumption of these foods. In the end, he was on a bovine-free diet for the most part, but occasionally ate his favourite chocolate snack bar and balanced the pleasure of eating it against the level of symptoms that consuming it produced. He no longer feared the sound of the lawnmower.

In this happy state, David took his family on a packaged summer holiday to a Spanish resort. The hotel they stayed in provided two meals each day and the evening meal centred on barbecued beef, out on the lawns. For the first few days, he watched his family enjoy the beefsteak, while he contented himself with the entrée dishes and the salads. Finally, the temptation was too great and, some way into the holiday, David succumbed to eating some of that beef. The next thing he was aware of was coming round in an ambulance, after suffering a suspected heart attack. In fact, it was a severe asthma attack.

This is the catch. If you avoid certain foodstuffs to avoid asthma symptoms, you may also increase your sensitivity to those foodstuffs as your body loses its adaptation to the challenge that such foods present. The subsequent consumption of the key causal foodstuffs can then result in dangerously powerful symptoms. This, by the way, is how elimination dieting works. When, after a period of abstinence, you return to a food you are intolerant to, you notice the more powerful effects it now has on your body.

The key in the above account is to separate the causes of asthma from the triggers of the symptoms of asthma. In the above case, antibodies to dairy and beef are the cause of the underlying asthma tendency, and newly cut grass aerosols are the normal trigger. I would suggest that the causal agents of asthma are mostly, if not all, components of infant formula milk. The triggers are mostly irritants of one kind or another. In cases like the one described, there may be an additional allergic sensitivity – for example, an allergy to grass terpenes, or an allergy to the faecal particles of dust mites – that increases the local levels of histamine to the point where the bovine immune complexes will tend to become drawn into the local tissues, thus bringing the complement reaction to those areas (the bronchiolar or nasal membranes).

For David, every meal that included bovine products was like a further immunisation to bovine antigens. When he ate the Spanish beef, antibodies that were ready to react to the meal would have been abundant in his circulation, and immune complexes would then have formed in abundance. Any trigger in the air would then have been able to induce the settlement of immune complexes and lead to a massive complement reaction in the tissues around his airways. The result would be a dangerous asthma attack.

A normal state for people with atopic symptoms is one of adaptive allergy. All the components are there to cause a severe allergic reaction, but the body produces higher levels of cortisone to quench that reaction. After a three- to eight-day period of excluding the problem food, the cortisone levels drop towards normal. Ingesting the problem food then generates the allergic reaction, but without the high levels of natural cortisone to protect the system.

Latterly, David consumes some foods of bovine origin, but lives a life free of the symptoms of asthma without the need for medication. The trigger of cut grass is always going to exist, but the absence, or extreme reduction, of antibody complexes in his system means that the asthma symptoms cannot now be expressed. He is in control and, more than twenty years on, still has no need of pharmaceuticals to control his asthma.

In recent scientific literature, the role of epigenetics has been invoked to explain why one identical twin may suffer from asthma while the other is free of such symptoms. However, a simpler and very plausible explanation would be that one child received some supplementary feeding with proprietary formula milk in the immediate postnatal period, while the other did not. Asthma is not a disease caused by genetic inheritance or air pollution. It occurred at much lower levels in the population just a couple of generations back, when industrial and residential air pollution from burning coal and coke was widespread and obvious.

Between one and five children had diagnosed asthma in each of the UK schools that I attended from 1952 onwards. Apparently, the figure is around one in five schoolchildren in the UK today. This equates to approximately six asthmatic children per classroom on average. Across Europe, asthma is now the most common chronic disease in childhood.

Eczema

The appearance of eczema early in a child's life is now commonplace. The use of cortisone-based treatments is now also commonplace and, given the ease of treatment, the cause of eczema seems unimportant. However, the same background process is going on behind eczema as is going on in asthma and arthritis, and atopic symptoms have been known to switch from one disease form to the other.

A component in the diet to which the immune system is reacting helps create circulating immune complexes. In the dermis, these immune complexes may adhere locally to produce eczema through a complement-type reaction that follows the settlement of antigen-antibody complexes. By this process, minor physical skin irritations can also turn into local extreme reactions such as wheal formation. Again, the most likely cure lies in diet modification, and it is common for the food types found in infant formula and weaning foods to be responsible for the symptoms.

Sometimes a chemical food additive also seems to play a role. Notoriously, tartrazine yellow (E102) was until quite recently (and had been for more than twenty years) a colourant of nearly all antihistamines that were available in tablet form. It is still used to colour certain medications and some Asian food ingredients such as 'hing yellow powder', a compound flavouring with asafoetida.

The safe, palliative treatment for eczema is the use of simple emollients, but a long-term cure requires changes to the diet. For most sufferers the use of cortisone cream seems far simpler. However, the continued reaction to foodstuffs and the accumulation of immune complexes will eventually bring about another set of symptoms, which are not so easily erased. Manufactured cortisone derivatives are instrumental in the masking of atopic reactions, but in use they also carry their own toxicity and risk of tissue damage.

Hay fever

I am writing this paragraph as the grass pollen season gets into full swing. Grass pollen is naturally prickly and irritating. It will irritate most noses if the levels of pollen in the air are high enough. So, what is the difference between an irritated nose and hay fever? The answer would seem to be a matter of degree. In hay fever, the inflammation of the nasal passages and the eyes is entirely out of proportion to the initial level of irritation. The initial irritation will cause the release of histamines, which itself causes immune complexes to gather and to induce a local complement reaction. This is all rather like the wheal reaction to scratching, but this time in and around the nasal passages.

Once again, dietary control is likely to bring about a long-term solution. If the bloodstream is free of high levels of newly formed immune complexes, then such a local complement reaction cannot be induced in this way.

Antibodies referred to as IgE antibodies play a significant role in immediate hypersensitivity and in facilitating the complement reaction. They associate with mast cells and help to trigger the release of components of complement reactions with these cells. Once established, IgE antibodies tend to play a continuing role in the inflammatory response to specific antigens in individuals who are atopically afflicted.

Rheumatoid arthritis

Again, this disease has a dietary cause and an underlying disease process similar to eczema and asthma. However, rheumatoid arthritis has a twist. Immune complexes resulting from immune reaction to food antigens are especially adherent to substances in the joints in this case, sometimes more so at lower temperatures, and thereby they cause the complement reaction to be focused on the components of the joint. This is a case of, 'Eat the food, feel the pain.' Dietary control can bring about rapid relief from rheumatoid arthritis symptoms, especially if previous steroid treatment has been avoided. All new inflammation can cease within a few days, and this happy state is followed by the gradual regeneration of the joints, where scar tissue does not inhibit this.

The varieties of arthritis would seem to depend upon the specific tissue substrate that the immune complexes tend to bind with. This might be a synovial component, a lubricant, a component exterior to the joint, or bone tissue in general. It may also be the case that the antigens from food, on finding their way into the bloodstream, are themselves capable of binding directly with target tissues before succumbing to antibody attack. This seems unlikely, though, given a probable superabundance of respective antibodies in the bloodstream.

Coeliac disease

This is a condition often associated with pains in the belly and a general run-down feeling. The disease is classically identified when there is damage to the villi of the duodenum linked to the consumption of food containing gluten. In some sufferers it makes life very painful and difficult as there is accompanying inflammation, and the consequent reduction in the active area of absorption of the gut restricts the uptake of vital nutrients.

It is believed that there are growing numbers of sufferers of coeliac disease in industrialised countries, and many of them are going undiagnosed. Symptoms tend to vary considerably in intensity and nature over time and have often caused problems in personal relationships, where one partner sees the other as dragging their feet and always complaining of feeling tired. My guess is that coeliac disease is commonly diagnosed simply as chronic fatigue syndrome before the relationship with gluten becomes apparent.

I have found that those with a diagnosis of coeliac disease commonly have a problem with the consumption of dairy products. I have a suspicion that dairy-based formula milk starts the damage, and makes the gut subsequently incompetent to handle gluten. There is often a clear pattern of inheritance of coeliac disease down the maternal line of a family. This is most probably due to the transfer of anti-bovine and/or anti-gluten antibodies and immune system components from mother to daughter, from one generation to the next.

Irritable bowel

The treatment of irritable bowel seems obvious enough. Dietary control should be based on the results of carefully structured elimination diets. The chief culprits appear to be bovine- and cereal-based foodstuffs. Pungent foods such as curries, which increase blood circulation locally, may assist with healing the bowel after abstaining from the offending foodstuffs. If a patient has recognisable IBS symptoms, they may find that elimination dieting gives very clear results, and allows them to quickly identify the causal food type.

Restless legs syndrome may be associated with irritable bowel. The bowel convulsions become linked to motor neurons through the autonomic nervous system and cause involuntary movement of the legs, which may be quite violent even during sleep.

Crohn's disease

There is probably a graduated series of gut disorders including irritable bowel, coeliac disease, and Crohn's disease, with Crohn's disease being one of the more painful and destructive. In Crohn's disease, the irritation of the bowel wall seems to draw in a complement-type reaction that causes pain, further damage to the gut wall, and the seepage of ingested antigens into the bloodstream, where a further range of anti-food antibodies is produced, making the disease very difficult to control. The disease is, therefore, likely to cause increasingly damaging inflammation of the gut wall the longer it is allowed to progress, and it is likely to prove more refractory to dietary control the longer it has persisted. However, dietary control could be very effective in the early stages. Again, the use of steroidal, anti-inflammatory treatment is likely to make later dietary control less effective. This is a direct effect of prolonged corticosteroid therapy.

The links between variant genes and Crohn's disease suggest a further dimension to the disease. In a very few individuals incompetent cell membrane proteins may exacerbate a situation caused by reactions to food intolerance. Or, inversely, reactions to food intolerance may follow from an alimentary canal lined with the genetic variant cells being affected by greater permeability.

Yet another new suggestion is that certain ingested bacteria begin to proliferate and change the behaviour of the E. coli bacteria already present in the gut. When the gut environment is changed by inflammation, it is to be expected that the local microbiome will be affected and that the behaviour of some bacteria within it may change.

It is at least possible that, in many such models of the disease, the initial damage to the gut wall is caused by immune complexes that have been generated by a reaction to dairy products. The damaged gut wall then provides a modified environment for the growth of new bacteria or fungi, or for changes in the behaviour of previously benign, established microflora. Such a change in behaviour from commensal microorganisms to facultative pathogens, when the host's physiology changes, is not without precedent. There is also the possibility that immune complexes formed in associated immune reactions may themselves be specifically adherent to components of the gut wall, which would thereby make a focused complement reaction possible.

Lower bowel cancer

Lower bowel cancer may be among the problems caused by food intolerance. There is no doubt that the gut wall becomes something of a battleground when the lumen contains foods that provoke an immune response, while blood capillaries in the gut wall contain the antibodies to components of such foodstuffs. It must also be relevant that patients diagnosed with

IBS are more likely to present later with bowel cancer. The surgical view is that carcinoma cells may propagate from polyps, which are generally subject to a biopsy when encountered. The origin of the polyps is presumably from an accelerated growth of the cells of the gut wall.

The link between food intolerance and cancer is readily observed through epidemiological studies, but the detailed mechanisms may yet require many years of research. It is possible that the consumption of dairy products, and/or beef, in atopic patients, will provoke the cell changes that prove so deadly (not directly, perhaps, but through a sequence of events from gut mucosal changes to gluten intolerance and to changes in microflora populations and inflammation, etc.). We now understand, somewhat, the role of *Helicobacter pylori* in the formation of gastric tumours, and we might anticipate that other microorganisms may similarly provoke changes in the cells of the colon lining and elsewhere.

In a recently published paper (zur Hausen, H., and Villiers, E.-M. de, 2015) the link is examined between different species and subspecies of cattle and the incidence of both colon cancer and breast cancer in the human populations that depend on them for milk and meat. This research could be extended to examine the species and variety of cattle from which the locally distributed infant formula milks originated. Nevertheless, the neonate reaction to formula milk will be somewhat specific to the source of the milk, which, in Europe, will be from the endemic cattle (*Bos taurus*).

In many parts of Asia, the cattle are largely of the zebu (*Bos indica*) species, though the infant formula consumed will likely be from European cattle (*B. taurus*). Where cross-breeding to improve herds has been carried out, the origins of cattle become a little less certain, but the higher the genetic input from *B. taurus*, the higher the incidence of observed breast and colonic cancers. Similar principles seem to apply to the cancer rates linked to the consumption of red meat, so that in Mongolia, for example, the consumption of red meat is high, but the incidence of linked cancers is low. This may be related to the genetic isolation of the cold-tolerant subspecies (*Bos taurus turano-mongolicus*) in that region, combined with a historically low rate of bottle-feeding of newborn children in Mongolia.

Osteoporosis

Often in the news, and often linked to dairy products in advertisements, osteoporosis would appear to be a diffuse reaction that may involve the complement system, but at a slow and steady rate within bone tissue. In other words, it may be that diffuse bone density loss predominates over bone rebuilding with the reintegration of calcium and magnesium. In one research project in working hospital nurses that directly linked milk intake to bone density, it was found that the highest milk intake was linked to lowest bone density. This was the inverse of the researchers' anticipated result. The implication is that this may be one of the least visible atopic diseases, but one with potentially severe consequences in life.

A possible model for the disease process is that consumption of, for example, dairy products, leads to the production of immune complexes that interfere selectively in the activities of osteoclasts and osteoblasts, or in their regulation. This may be by the depletion of local bone growth-promoting metabolites, or by the depletion of growth hormones. Levels of dietary calcium do not seem to affect the course of the disease to the extent that might be anticipated, with or without adequate vitamin D. Osteoporosis is a known risk in coeliac disease, and is assumed to be due to poor calcium absorption. It may equally be a consequence of the immune disruption of the restructuring of normal bone tissue.

There is a likely relationship between osteoporosis and spinal deformations such as scoliosis. Curvature of the spine would appear to result from the differential loss of bone density or the differential bone regrowth during the constant recycling of the components of bone tissue.

Hypertension

Hypertension has a multitude of causes, but is included here because atopy cannot be disregarded as a principal cause. When a newborn infant drinks formula milk, its gut is far from completely mature. The 'milk' can be transported with little change into the bloodstream and around the body, because it is physically and chemically somewhat like human milk. This brings a wide range of tissues into direct contact with the feed components, and a degree of binding between the two may result. This would include the lining, or intima, of the blood vessels.

In arterial disease, cholesterol-rich layers are produced in the arterial intima to smooth over areas of inflammation of the artery lining. This would appear to be a similar process to atopic eczema, but one affecting the intima of blood vessels. Alternatively, it could simply be that the vessels become damaged by a diffuse complement reaction series while transporting antigen-antibody complexes around the body. In either case, cholesterol would seem to be produced as a biological repair or protection compound. It is probably the continued irritation and inflammation of the intima that destabilises the established plaques and turns them into potentially fatal vascular hazards.

It is possible that some immune complexes adhere selectively to the intima of arteries and attract the complement series of reactions, resulting in the localised release of proteolytic enzymes. While these processes are going on, the repairing of the vessel will include the increased production of the collagen fibres that serve to reduce the elasticity of the blood vessel wall and reduce the maximum internal bore of such a blood vessel.

In old age, the range and multiplicity of food-antibody complexes in the bloodstream may cause raised complement proteolytic enzyme levels. This might explain why the protective layers of cholesterol-rich cells are produced in greater quantity at constrictions in blood vessels or in regions of higher turbulence of blood flow. In these locations, the turbulence would tend to bring, over time, quantitatively more complement proteolytic enzymes into proximity with the intima.

While there is little doubt that the past consumption of modified fats in processed foods had a worsening effect upon fatty arterial deposits, the likelihood is that most arterial wall damage now has atopic origins. More research focus on bovine and other allergenic foodstuffs in arterial disease might produce breakthrough results. Certainly, some studies have indicated that atherosclerosis can start in teenage children and this seems to parallel the occurrence of, say, rheumatoid arthritis in some very young children. After several generations, during which food intolerance reactions have been passed down the maternal line, along with bottle-feeding and regular consumption of dairy products through the generations, it is not surprising that young children begin to exhibit disease processes that were previously found almost exclusively in adults.

Medical science is beginning to appreciate the role of food allergies in the development of hypertension, and simple observation has demonstrated a rise in blood pressure on exposure to certain foods in those so affected. The more pernicious impact of food intolerance, generally, has yet to be fully explored in relation to hypertension. A study linking childhood atopy and adult hypertension might reveal some interesting statistics.

After causing a variety of symptoms to occur, the atopy finally exerts its impact on internal structures such as the blood vascular system, where the long-term effects of carrying food-antibody complexes around the body cause arterial and cardiac damage, expressed at some point as hypertension. Some epidemiological studies have already linked external symptoms such as psoriasis to internal conditions such as hypertension and ischaemic heart disease.

Episodic hypertension may relate directly to food intolerance or to exposure to allergens in those with existing atopic disease. Exposure to the causal foodstuff, or allergen, causes extreme raised blood pressure, which may persist for several hours or days before returning to normal. The seasonal release of specific pollens or fungal spores may trigger episodes of hypertension in some individuals.

Early-onset diabetes mellitus

It had long been known that in early-onset diabetes there was inflammation and destruction of the islets of Langerhans cells of the pancreas. This process was labelled as an autoimmune disease. It took a recent study of indigenous North Americans to demonstrate that bottle-feeding could be directly linked to an increase in early-onset diabetes among these people. The mechanism would appear to be similar to other atopic disorders. But in this case, chance may lead to immune complexes being selectively adherent to the components of the islets of Langerhans cells, and may bring about their destruction through the complement cascade reaction.

Early-onset diabetes mellitus is on the rise in many parts of Europe and, though some cases may be identifiably of genetic origin, the recent rapid increase must reflect an environmental cause. It most probably relates to feeding babies infant formula. That type 1 diabetes is commonly associated with atherosclerosis suggests a connection with the general atopic pattern of disease symptoms. It may well be the case for this disease, as with some other atopic symptoms, that the offspring of a mother or of a maternal grandmother who was affected by bottle-feeding are suffering now. Cross-feeding of the newborn in an earlier generation provides for a wider range of food antibodies in subsequent generations (through the maternal line), by the transplacental passage of antibodies and immune system components from one generation to the next, during gestation, and by the provision of antibody traces during lactation.

I am not aware of any experimental data linking dietary change to the recovery of insulin production in affected individuals, though this should be investigated. There are examples of people being cured through stem cell transplantation (see the section on multiple sclerosis below), which would indicate that dietary change might be simpler and cheaper. The early implementation of dietary change, based upon measured levels of anti-bovine antibodies, would be paramount to bringing about any possible recovery. I would consider foods of bovine origin as the initial primary suspects, because it is my own observation that early-onset diabetes tends to occur in families where sibling children, if any, have a history of intolerance to such foods. This is also indicated by the results of research investigations upon indigenous North Americans.

I believe the label 'autoimmune diabetes' should be reserved for the rare instances of genetically linked early-onset diabetes. In most cases, it will be an immune system corrupted by cross-feeding of neonates, or by maternal antibodies passed on *in utero*, that destroys the potential to produce insulin. In such a case the use of the term 'autoimmune diabetes' is misleading. The word 'autoimmune' should be restricted to cases where the corruption of the immune response by production of antibodies to foodstuffs, and the resultant complement reaction, has been ruled out by thorough scientific investigation. The current use of the term 'autoimmune' simply obscures the origin of what are mostly atopic disease symptoms.

In relation to late-onset diabetes, there is the possibility of a perfect storm of metabolic disruption. I have observed that many late-onset diabetes sufferers have an intolerance to dairy products. It is likely that the dairy intolerance makes them eat foods of bovine origin addictively through the action of cortisol on their mood, along with the sugars and starches that usually accompany such foods helping to raise their body mass index.

Overeating stresses the production of insulin. Immune complexes arising from reaction to foods of bovine origin may become focused, to a degree, on the islets of Langerhans, as they do in early-onset diabetes. The result might be an acceleration to type I diabetes. A piece of useful advice to late-onset diabetes sufferers might be to reduce the amount of food of bovine origin in their diets, and they should reduce this to none, if possible, if they are already aware of their sensitivity to dairy products. In any case, it would do little harm if they then rebalance their diet to make up for the loss of dairy foods.

Multiple sclerosis

Multiple sclerosis, in some ways, fits the pattern of atopic disease. It retains its own mystery, simply because the observed effects can be so horrendous. Sufferers are very much victims who are at the mercy of a disease that jerks them around like a puppet on strings, through bad attacks and apparent remissions, and then on to more severe symptoms. This pattern of variability suggests there may be some environmental component to the disease, possibly with a variable lag period in the response to those environmental factors. Fluctuating levels of food antigen-antibody complexes may affect the rate of destruction of the myelin sheaths around the nerves through complement reactions.

The initial damage may be self-limiting in some way. The recent use of stem cells in lymphocyte replacement therapy fits in with the above observations. Autologous hematopoietic stem cell transplantation (AHSCT), which is already in use for multiple sclerosis, lupus, rheumatoid arthritis, and type 1 diabetes, is, however, a drastic and expensive treatment. AHSCT involves the harvesting and the *in vitro* culturing of a patient's bone marrow stem cells and reintroducing them to the patient's circulation following chemotherapy to destroy any existing mature circulating lymphocytes. Nevertheless, AHSCT is producing some remarkable recoveries from the diseases listed above. It is a process that effectively resets the immune system to a significant degree.

Myalgic encephalomyelitis (ME)

Myalgic encephalomyelitis has been used as a descriptive term for a syndrome that may have many different causes. It is essentially a collection of symptoms that are believed by some to be caused by the virus administered to induce immunity to poliomyelitis, or a mutant thereof.

The definition of ME has become obscured by a tendency to describe any illness with similar symptoms to ME, rather than keeping the reference to the attenuated poliovirus. For some of the sufferers of ME whom I have known, simple attention to diet, with exclusion of dairy or wheat products, for example, has brought about dramatic relief from symptoms that had been diagnosed as ME.

Autism

Some types of autism are inherited genetically, but there is plenty of evidence to show that, where careful dietary modification has been tried, there has been improvement for a very high proportion of autism sufferers. As part of the disease process, the involvement of food-triggered immune complexes in the destruction, or incapacitation, of neurotransmitter precursors, or even in direct brain cell interaction, seems possible. Autistic children are commonly born to mothers with existing atopic symptoms. It appears to be at least a second-generation atopic disease. It may be the result of an increase in the permeability of the placenta, caused by atopy in the mother. There are reports of the presence of anti-brain antibodies passed from the maternal circulation to the foetal circulation.

Autism is, perhaps, one of the extreme dissociative disorders in its impact on the sufferer and on the sufferer's family. The rapid improvement and total social change that a dietary approach can bring about makes one wonder why this approach is not more widely attempted. The only drawback in the dietary approach is that it can be quite complex and drawn-out before the optimum diet is established. Otherwise, it is the most certain and, usually, the most economical approach. It is also entirely drug-free. Is it because there is no apparent profit for pharmaceutical companies that this approach is one of the least scientifically researched?

Autistic children, as a group, are very capable of persuading their parental carers to provide them with the very foods (foods to which they seem to be addicted) that cause their continued autistic behaviours.

Reading various guides to dietary control of autism led me to see that autism is just another affliction caused by the cross-feeding of infants, though it tends to appear in the next generation. Experimenters obtaining good results from the dietary control of autistic symptoms have tended to remove gluten and casein from the diet. This observation provides a further link to the diseases more commonly accepted as atopic. Cereal grains and cow's milk are two foods that have entered the human infant diet relatively recently.

The action of gluten in autism has been described as similar to the binding of opiates with brain receptors. What is it that makes children susceptible to the triggers of autism? It must usually result from a corruption of the maternal immune system that leads to the passage of unhelpful antibodies to the infant. These acquired antibodies may help to create a leaky gut that allows antigen-antibody complexes to form and to migrate through the blood vascular system to the brain. In many children, brain function is normal up to a time when there is a sudden dramatic change between the age of one and ten years.

The MMR vaccine as a trigger

It is very likely that all commentators and researchers on the potential link between the MMR vaccine and autism, including Dr Andrew Wakefield, are generally correct. It is possible to envisage that the dietary element in autism is due to the increased permeability of the gut in atopy sufferers. It is further possible to envisage that, in children carrying antibodies to common food groups, there already exists a tension between the gut content and the immunocytes and antibodies circulating in gut blood vessels.

The use of an immunising jab that may cause some localised gut wall inflammation could be just the trigger required to bring food antigens and circulating antibodies into closer proximity. This would be enough to trigger more extensive damage to the gut wall and to increase its permeability. In some cases, the upshot might be the beginning of the disease processes that lead to the symptoms of autism.

The key point in this scenario is that the body is probably on the verge of receiving the critical minor gut damage that triggers symptoms. If the MMR jab failed in causing this trigger, the next enterovirus or other gut invader probably would. Therefore, the effect of the MMR vaccine is significant insofar as it may bring inevitable symptoms forward in time (by a small margin). Nevertheless, in this scenario the MMR vaccine has probably precipitated autism in someone with a predisposition to the disease. The statistic that emerges from these scenarios is that the MMR jab would not significantly increase the number of cases of autism within a population, but it may bring forward the occurrence of autism in some of those children who are predisposed to develop the condition.

In cases where it is certain that the MMR vaccine is not the trigger, for example, when the MMR jab is not given, then it is possible that the attenuated measles virus is acquired from another source or that some other infection or incident has damaged the integrity of the gut wall. The spread of a vaccine virus within a population may bring about unexpected consequences.

My own observation is that, in most cases of autism in a child, the mother is suffering atopic symptoms of one form or another – typically asthma or eczema, but sometimes migraine or arthritis. My guess is that this may also be true for many mothers of those young narcoleptic patients whose condition developed immediately after immunisation with an influenza vaccine.

Alzheimer's disease

Current research indicates a strong link between Alzheimer's disease and diet, especially with regard to fats. The suggestion is that some dietary fats tend to increase the formation of amyloid protein plaques in the brain. There are also observations that link Alzheimer's to a range of atopic symptoms. One commentator likens the formation of amyloid plaques in the brain to the formation of cholesterol in blood vessels. It may well be that amyloid protein is produced when complement components, or other immune entities, enter the brain. However, it remains conjecture as to whether the amyloid is protective, or the central cause of destruction of brain function, or whether it is just a structural infill after nerve cell death.

Various studies have identified an antimicrobial property in amyloid protein. Yet other studies reveal a connection between the occurrence of many of the diseases described in this book and the occurrence of Alzheimer's disease. The suspicion must be that Alzheimer's is one more example of the atopic range of symptoms that normally become apparent later in life. This would explain the rapid rise in the identification of Alzheimer's disease in patients who are at the lower end of the normal age range for symptoms of senility. Alzheimer's and autism are linked by the apparent common involvement of raised amyloid protein levels, but amyloid protein is likely to be just a filler that works to maintain the structural integrity of damaged tissues.

A strong association between arterial disease and Alzheimer's suggests that localised ischaemia in the brain may be the cause of nerve cell degeneration, which then leads to amyloid deposition as neuronal activity decays. This fits with all the observations on atopic disease symptoms discussed in this book. In addition, as arterial disease becomes ever more widespread, so Alzheimer's disease may affect a greater proportion of the population.

One thing is certain: the current approach to Alzheimer's must evolve to become preventative and curative rather than being a chemical-based palliative process. In recent years the chemical treatment of Alzheimer's patients in the UK with antipsychotic drugs has been responsible for an excessive early death rate. This is highly unethical, especially in a situation where the patient is not reasonably able to give valid consent to the treatment, and is equally unable to be willingly involved in an assessment of the effectiveness of such treatment.

Prostate cancer

There are known nutritional links to prostate cancer, but it does seem to me that the prostate mucosal tissue may very well be a focus for the selective adhesion of food-antibody complexes. Immune complexes can trigger the start of the complement series of reactions, which causes inflammation, irritation – and, subsequently, perhaps – leads to cancer within the gland. Increase in gland size is known to be merely indicative of an increased risk of cancerous growth, though not diagnostic.

I think researchers might, usefully, look at the correlation between increased prostatic size and atopy, and any subsequent increase in the likelihood of cancerous growth. The functional regions of the prostate and of the mammary glands are both mucosal tissues that are prone to malignancy. Within populations having no dairy industry, breast cancer and prostate cancer are both rare. They are certainly well below the levels in the West. In China, the occurrence of breast cancer is now rising rapidly in urban populations but remains low in the rural population, where dairy products are still largely shunned and where more babies tend to be nursed naturally by their mothers.

Atopy mechanisms: an overview

It appears that many of the more severe atopic symptoms appear in a second or third generation of the maternal line. Acute hypersensitivity to foodstuffs like peanuts or fish or to rubber latex, and some of the more severe psychological changes, can usually be traced back to an atopic mother and, in some cases, an atopic grandmother. In the case of latex reactions this can often be traced back to the feeding bottles with natural rubber teats used by previous generations.

If you feed a newborn baby girl with an alternative to her mother's milk you risk changing her life and the lives of all the succeeding generations down that new maternal line.

The effects of cross-feeding of neonates appear to be largely irreversible. This does not only affect humans but also most mammals, including all domesticated animals, animals that have been rescued from the wild, and zoo-bred animals. In the case of, say, abandoned seal pups, cross-feeding may be of less consequence if the bottle feed is free of ingredients based on marine food sources that the seal might feed upon in later life.

In the case of mammals such as lions that prey on ungulate mammals, there is the possibility that they may develop food intolerance reactions to any cattle (or even deer) prey, if fed as cubs on a formula based on bovine milk. In many zoos they are fed on bovine meat. More importantly, females so raised would never be fit for being returned to the wild, as they would bring atopic disease into the wild population – a consequence that would nullify efforts to conserve a fully natural and wild population.

It is likely that the extra burden of atopic disease would lead to the eventual elimination of any atopic maternal line by natural selection – the ultimate determinant of fitness for survival. Similarly, for pet animals that are often fed complex mixtures of foodstuffs, the use of substitute milk to feed neonates can result in a range of atopic symptoms in adult animals that closely mirror those of humankind: eczema, arthritis, heart disease, etc. – and, possibly, some of the mental illnesses that result from atopy.

Variable properties

The variable properties of immune complexes (specifically, the complexes formed as the result of a reaction between a food antigen and the specific antibodies to that antigen) are partly a result of the genetically determined structure of the antibody, which will affect the physical behaviour of the combined complex in terms of specific adhesivity, sensitivity to pH, and temperature, etc. In other words, the consequences of the formation of food-antibody complexes are largely determined by chance. Cellular surface antigens within the body will also have a structure that is genetically determined, and are therefore also somewhat variable.

In one person, wheat-specific antibody complexes may prove relatively harmless. In another person, an antibody produced in response to wheat may result in immune complexes that adhere specifically to joint linings, resulting in arthritis. If the response is to a component of a plant, there may be a similar reaction to the whole plant family. In the author's case, there is an arthritic reaction when any species of the Poaceae (Gramineae) or related plant families are eaten. These include wheat, rice, maize, rye, and sugar cane, among others, though the reaction can clear within a few days when eating them is then avoided.

This apparently random behaviour of the immune system is an indication of the complexity of analysis required when designing or interpreting experimental models. The fact that inherited aspects of food intolerance reactions may be determined by two distinct and separate inheritance systems, the genetic and the maternal antibody transfer processes, should always be kept in mind. Recent research indicates that the maternal antibody transfer also encompasses some degree of immune system programming in the recipient that takes it somewhat beyond the accepted concept of passive transfer.

Adhesion/adsorption and metabolite depletion: a potential mechanism

The first theoretical model that fitted clearly with observations on my arthritis was the adhesion of immune complexes to joint tissue components following the consumption of specific food types, resulting in arthritic inflammation of the joint through subsequent complement reactions.

Immune complexes are formed by the reaction of immune system antibodies with ingested antigens to form antigen/antibody complexes. Presumably, the tissues of my joints have some substance or substances which have a surface structure that traps or adheres to the immune complexes of specific shapes and sizes. Thus, the response may be specific to a single antigen type with its conjoined antibodies. I have found that arthritic pain can be felt within twenty-four hours of consuming the offending foods.

Considering this model led me to the concept of immune complexes in general being selectively attracted to a specific surface feature of one of a wide range of metabolites and substrates. This provides a widened theoretical model that encompasses not only the macro-inflammatory responses of arthritis, asthma, hay fever, eczema, etc., but also the micro-adsorption or chelating reactions that could remove or damage metabolites within the bloodstream and elsewhere. A putative explanation for such atopic diseases as autism, depression, and a wide range of such metabolite depletion diseases is now possible. If neurotransmitter building blocks were scavenged by chelation in this way from the bloodstream or from nerve tissue, the processes of the central nervous system could become blocked or unbalanced.

The immune system has evolved to react safely with the range of antigens that it has naturally faced. What is now so different is that the new antigens faced by the immune systems of neonates include components of such recent foods (in terms of human existence) as cow's milk and cereals, especially gluten-rich cereals.

In affected individuals, large amounts of immune complexes may be formed with cereal antigens after each meal. These foods have been part of the neonate diet for a relatively short period of human existence. Following the development of infant formula feeds and wheat rusks, both have been introduced as food before the infant gut may have fully matured to act as a barrier to the passage of such complex molecules. It should not be surprising if a population that has been raised with such novel nutrition were to begin to experience novel disease symptoms. The normal mechanisms for the breakdown and removal of immune complexes are also likely to be overwhelmed.

There was also the historical approach to baby feeding based on pap, or bread soaked in milk and/or water, and this may have had a major influence on child survival. When a mother who has been raised using such a novel system of nutrition becomes pregnant herself, the foetus is raised in a novel uterine environment where the placental connection will allow a degree of free passage to some of the new antibody complexes carried in the mother's bloodstream. Some of these components may damage the placental barrier with the complement reactions they induce and allow other molecular species free access to the foetal circulation and organ systems. We could refer to this as 'leaky placenta syndrome'.

Towards the end of gestation, a major transfer of maternal antibodies and some maternal immunocytes to the foetal circulation occurs, conferring on the infant the full range of food intolerance capabilities that its mother has developed. The transgenerational passage of atopic disease through the maternal line is thus ensured. There may also be an explanation here for premature birth that is not the result of rhesus-negative genotypes.

Food-related mental disturbance has long been a topic that challenges the orthodoxy of Freudian models. Instead of psychosomatic disorders, we should think of 'somatopsychic' illness, as some authors have described it. Despite the widespread acceptance of Freudian thinking, over many years, there is something more satisfactory in the somatopsychic approach.

The Freudian approach is dependent on emotional binding for acceptance. The somatopsychic approach makes no such demands: every patient becomes a victim of environmental factors and of chance. All of us have feelings that are the product of our upbringing and experience, but only a few of us employ these to explain extraordinary behaviour. Symptom-related psychoanalysis should be replaced by elimination dieting to find the trigger factors for many a mental health condition. Metabolite depletion could affect many body regulatory systems, but the somatopsychic illness model might provide an example for research.

In his book *Not All in the Mind*, Dr Richard Mackarness describes his encounter with a patient who was due to have brain surgery to break a pattern of self-harm. By changing the patient's diet, the pattern of depression and self-harming was stopped until a time when the patient, with new-found confidence, returned to her old eating and coffee-drinking pattern. Her self-harming tendency returned, but was curtailed by a guided return to her safe diet.

Early-onset diabetes can be viewed as an example of a disease linked to metabolite depletion. In this case, insulin is the metabolite, but the depletion occurs through the destruction of the islets of Langerhans in the pancreas. Depending on the system affected, repeated or continuous metabolite depletion – through atopy – could put immediate and long-term stress on the body. Control mechanisms could be put into overdrive, and system failure could follow. Think of the effect of chelating growth hormones or other endocrine system metabolites. The range of systems and the potential impact of such metabolite depletion could be considerable.

It is very likely, where metabolite depletion has begun, that there would also be other localisations of the immune system complement reaction. Vascular disease and osteoporosis are very likely to be examples of a more diffuse mediated reaction of the complement system. But in osteoporosis, for example, a metabolite depletion reaction may also occur which removes the bone morphogenetic proteins or other factors that favour osteoblast (bone-building cell) activity. In some cases of osteoporosis there may be a direct action of immune complexes within bone tissues. Or there could be a controlling adsorption interaction of the immune complexes with the metabolites that control bone growth.

In either case it is essential to track down the root cause of the immune complex production as early in life as possible, so that the bone might have a chance to regenerate and rebuild its density. Without careful dietary modification, the disease normally continues with further loss of bone density. The avoidance of provocative food groups may be more useful than the addition of ever higher doses of mineral and vitamin supplements.

Reversible destruction

It has been my observation over thirty-five or so years that, where an atopic degenerative disease is truly stopped by dietary control, affected joint membranes, metabolic systems, and, indeed, somatopsychic symptoms are self-healing. If rheumatoid arthritis is stopped by the correct dietary changes, then the joints will substantially repair and rebuild. If the symptoms of rheumatoid arthritis are simply controlled by NSAIDs, together with modified steroids, then the joints do not rebuild and repair. Misshapen hands remain misshapen.

Recorded clues about the importance of diet in mental illness came from Holland in the Second World War, when civilians were deprived of normal foodstuffs by the occupying

German army. The staff of mental hospitals who stayed with their patients noticed that the schizophrenics lost their schizophrenia when they had only weeds, berries, and flower bulbs for sustenance. When the Allies managed to get normal rations to them again, the unfortunate patients returned to their schizophrenic state. This is a clear example of the principle of reversible somatopsychic illness.

Potential research areas

It is beginning to appear likely to the author that a key component in inflammatory atopy is the basal lamina of affected tissues. The role of the basal lamina may well include the organisation of differentiating cells within respective tissues. The basal lamina is also a product of those cells and has specific properties according to the type of cell that assembles it. It has also been demonstrated for some muscle cells that antibodies can be produced that become bound to just the basal lamina around synapses. The components of basal laminae are type IV collagen and various complex glycoproteins and proteins. This makes the basal lamina an ideal target for immune complexes with milk or cereal proteins, which will be relatively new molecular structures to the body (in evolutionary terms).

Locally induced inflammation, in which local irritants cause the release of histamine, may be sufficient to bring any circulating antigen-antibody immune complexes that are food-related into contact with the basal lamina components. These complexes will trigger a major local allergic response. In the case of asthma, an attack is usually initiated by the inhalation of a known irritant. Through the complement reaction, immune complexes from ingested foods of bovine origin are then able to trigger a major allergic response locally in the bronchioles. This correlates with a published description[1] of the pathology of asthma that there is 'fragility of the airway surface epithelium and thickening of epithelial reticular basement membrane with the presence of an inflammatory infiltrate, comprising activated T lymphocytes together with activated and discharging eosinophils'. - A scientist's description of the complement reaction at work.

Antihistamines

One observation that deserves to be repeated is that antihistamines, for more than twenty years, were sold in the UK in tablet form, coloured yellow with tartrazine. Many who took these tablets had their symptoms worsen or came out in hives, despite the presence of the antihistamine. One could almost use the example of having a reaction to tartrazine as a marker for atopic illness. In young children it may be a key component in the development of some somatopsychic symptoms.

Antihistamines are normally quite effective in the short term, but they simply hide the symptoms of the underlying disease process. They do not control atopic illness. In the long term the side effects of the antihistamines become more apparent or the response to the drug becomes very limited. That said, many hay fever sufferers do manage to control their symptoms from year to year with simple antihistamines (without the tartrazine) and are grateful for the relief.

---

[1] (Br Med Bull. 1992 Jan;48(1):23-39. Pathology of asthma. Jeffery PK)

The correlations between atopy in early life and disease processes in later life deserve serious investigation. The early eczematous rash behind the knees and inside the elbows may turn out to be the portent of many of the diseases that make old age so uncomfortable for so many in our culture, and may also be linked to many of the symptoms that afflict people throughout their lives.

Anti-inflammatory drugs

If you are in agony and need fast relief from inflammatory pain, then steroids or NSAIDs may well be the first choice. They are so effective though, that there is a strong temptation to also employ them for chronic conditions.

This is usually a mistake. A major problem with anti-inflammatory drugs is that they mask symptoms so well that it later becomes difficult to investigate the true cause and nature of the inflammatory response and difficult to identify the causal agents. For sports injuries the cause may be clear-cut but, for arthritic or rheumatic pain, the cause may be effectively masked and unsuspected.

Statins, anticoagulants, etc.

These lifesavers are also not always without serious drawbacks. For acute conditions they are truly lifesavers, but the long-term use of some anticoagulants can lead to muscle wasting and other side effects. The use of warfarin as a blood thinner makes the use of other alternative therapies dangerous (though normally harmless), especially where patients are on diets low in vitamin K.

Again, the use of these pharmaceuticals tends to generate a feeling that the complete disease has been held at bay, when in fact only one or two symptoms are being suppressed. These types of drug tend to be used without much consideration about the cause of the illness: all focus is on the symptoms. There is a common assumption that genetics and lifestyle are the key issues in human health, even though many studies indicate alternative explanations.

Atopy epidemiology

There seem to be reasonable grounds to link all atopic diseases to a similar basic cause – the inappropriate feeding of neonates on milk from other species. The symptoms of atopy produced would appear to depend on the nature of the foodstuffs fed to neonates, on the genetically inherited structure of antibodies and other molecular structures, and on the competence of the full immune system. Nevertheless, there is a direct connection between, say, asthma and eczema, or colitis and hay fever, and a connection shared by other atopic symptoms. In addition, almost any increasingly common atopic disease, such as Alzheimer's disease or autism, might be linked back to the bottle-feeding of neonates.

Initially, this common causality seems very unlikely, but, when the complexity of immune response is considered, it begins to seem more plausible. And, for many diseases, the concept of metabolite depletion – or receptor blocking, through selective adsorption or adhesion – provides a very credible explanation of multiple symptoms through the interaction of antigen-antibody complexes. Thus, the early diet can affect almost any aspect of life, from the skin to the deepest part of the circulatory system or areas of brain tissue. A difficulty only arises from the perception that immunity is a strongly controlled process and that the chance attraction of immune complexes to species of surface proteins and glycoproteins, wherever they occur in the body, seems to be too serendipitous.

Studies of the health of recently migrated communities indicate that there are very large changes in the types and levels of various illnesses, including cancer, which appear to result from small changes in neonate diet and lifestyle. More westernised communities tend to have a higher incidence of lower bowel cancer, for example. The suspicion must be that changes in infant nutrition, which have been introduced under pressure from paediatric practitioners, have had a part to play in these epidemiological changes.

Such practitioners have changed over time from being advisers to being police, judge, and jury, as many new parents discover when raising an infant does not run as smoothly as anticipated. Consequently, their impact upon newly arrived communities can be dramatic. They have had the power to inflict bottle-feeding on infants who are not meeting the target growth, according to tables that may have been derived from the growth patterns of bottle-fed babies.

Diet-related psychotropic effects

Hyperactivity in childhood
Most hyperactivity in children appears to link back to dairy-related atopy. One aspect of atopic illness is that it seems to make certain individuals more prone to reaction to the toxic additives that are still a component of many manufactured foodstuffs. When nearly all antihistamines were coloured with tartrazine yellow, a patient would attend a doctor's surgery with some rash or irritation, for which antihistamines would be prescribed. The condition would then become much worse.

As one author put it, the patients were a self-selecting group with a predictable reaction to tartrazine yellow. We ought to take note that a substance that can cause a reaction resulting in hives can also cause a profound change in behaviour. This is an indication of the reality of somatopsychic illness in daily life.

The behaviour of children is complex and driven by a multitude of modifiers, both social and environmental. In childhood, it is commonly observed that the consumption of artificial colorants and, sometimes flavourings, results in an extreme modification of child behaviour. My observation is that the more extreme the behavioural modification, the more evident are other atopic disease symptoms such as eczema, glue ear, or asthma. The consequences for the child are often profound in terms of educational and social achievement. Parents are often surprised at the intellectual capacity of such children once they reach the developmental stage of being able to control their behaviour somewhat, through the development of self-awareness.

The ideas linking sugar intake with hyperactivity are probably related to the observation of children's behaviour at parties, where sugar-rich foods are consumed and their behaviour then becomes a bit wild. Sugars tend to be associated with artificial colours and flavours that may contribute to hyperactivity. The influence of the child group as a behaviour modifier is probably also significant in this instance. Furthermore, there may be a role played by the yeast candida, which flourishes in a sugar-rich environment. An increase in growth of candida within the gut may follow damage to the gut wall, which has been caused by anti-bovine antibodies.

Hyperactivity in adulthood

Hyperactivity is a symptom of the dissociation between awareness of an activity and control of that activity. Children who turn into monsters after ingesting certain foodstuffs are aware of what they are doing but are not in control of their actions. It is as if moral references have been switched off or neutralised. After a period of uncontrolled aggression, it is difficult for the child to apologise for actions that somehow happened but were beyond the child's control. It is as if all thought of fault or guilt is extinguished for a while, returning only as the somatopsychic effects of foods or additives wear off.

Adults are generally a little more sophisticated in their actions than children are, and able to spin excuses and to hide the reasons for their actions more skilfully. Nevertheless, there are occasions when adults appear to suffer the impact of the same food additives in a similar way to children.

During the great miners' strike, on 30 November 1984, there was a chilling event in which two miners – who were described in court as being of otherwise good behaviour – dropped a twenty-one kilogram lump of concrete from a motorway bridge onto a taxi passing on the motorway below and killed the driver, David Wilkie. I am not in possession of all the facts in this case, but it led me to wonder about the reality of adult hyperactivity, sometimes labelled 'dissociative disorder'. Could it be that striking miners, living on donated snacks, were eating excessive amounts of specific foods or food additives and suffering the psychotropic consequences that might have amplified the group behaviour tendency?

Depression

Many illnesses relate to depression, ranging from seasonal affective disorder (SAD) to bipolar disorder or manic depression. The explanations of these illnesses are wide in range as well. Those with SAD sometimes respond well to increased light levels, natural or artificial, and low light levels in winter and autumn are blamed for their depression. What perhaps should be considered more deeply is why only some people are affected in this way and others remain entirely unaffected. It may be genetic, but in most of the lands where SAD affects significant numbers of people, such as Sweden and Finland, the total consumption of milk products is particularly high, so it might also be linked to this diet.

Psychology writers have concluded that it is because life has become so detached from nature in the West that people have become disillusioned and depressed, and that they are suffering because otherwise life in an industrialised society would be too stimulating. This seems to follow Freudian thinking to an extent, and embodies an element of guilt and blame.

If atopic illness is the true cause, then we may have to make a further adjustment to our thinking, and perhaps dismiss some of the key aspects of Freud's postulates. People suffering from depression are often users of cortisone-based pharmaceuticals. One of the obscure side effects of cortisone treatment is that it exacerbates depression and tends to add mania and paranoia to the mix. The asthma inhaler, or the skin cream for psoriasis, can bring about unexpected mental consequences. This type of side effect may occur in up to 20 per cent or more of the users of some pharmaceutical cortisone preparations.

One of the questions often ignored when discussion focuses on the symptoms of stress is, 'Why should some folk suffer symptom A under stress, when others suffer symptom B, and yet others suffer symptom C?' In other words, what is it that predisposes people to express their response to stress in one specific manner?

Headache, irritable bowel, fatigue, or depression, may each be attributed to stress without any consideration of the mechanism of predisposition to the specific symptom. This helps to conceal the possibility of a dietary path to the resolution of the disease process. Many symptoms can be related to atopy, and so a diagnosis of stress may be a further indication of the wider impact of atopic illness. This increasing spread of atopic psychosis results from the widespread cross-feeding of newborn infants and is not 'a rational Freudian response to industrialised society'.

There are numerous accounts in a variety of publications describing near miraculous recovery from various psychoses through dietary modification. On a more mundane level there are many who modify their diet on the advice of alternative practitioners. The symptoms that they seek to resolve may relate to physical symptoms such as enteric, musculoskeletal, or skin problems. In the resolution of these symptoms, they may also recognise a release from depression, the existence of which they may not previously have acknowledged.

Obesity

Obesity has been in the news for a long time. For a time, it was seen from the UK as a result of living an American lifestyle, until detailed investigations on UK obesity levels were highlighted by the media. In all the hype, obesity is nearly always portrayed as the cause of so many life-threatening disorders such as cancer and heart disease. There are very few reports that attribute the status of 'symptom' to obesity or describe obesity as caused by a disease process that is also linked to such problems as heart disease and cancer. It has been assumed that obesity is self-evidently the root cause of these ailments.

I believe that this assumption is the result of simplistic reasoning. Consciously or subconsciously, obesity is associated with laziness and gluttony. The possibility of anyone suffering from a compulsive food addiction is ignored because of the need to place blame on those suffering from obesity. In truth, obesity is generally an atopic symptom. The action of antigen-antibody complexes from foods includes the reduction of circulating hormones, such as leptin, that promote the feeling of satiety. The result is often that food intake is controlled instead by emotion. Simultaneously, damage is being caused to blood vessels and other body tissues by atopic processes. It then appears that obesity has caused this other damage, rather than both being caused by the reaction of the immune system to foods, usually foods of bovine origin.

The current attempts to limit the calorific content of ready meals are unlikely to resolve the issue of obesity. Those who feel the need to eat excessively will simply buy more than one portion or serving. There are no such simplistic solutions. The high price of addictive drugs does not preclude acquisition of a needed dose; however it might be obtained.

Obesity occurs when weight gain goes out of control. Most of us experience periods of stress when we tend to lose weight and periods of relaxation when we tend to get back into condition, but more and more people are losing control of the amount they eat. Yes, this is a time of plenty and yes, it has never been easier to eat more calories than you need. However, the new phenomenon is actually an addiction to eating. It is very much a symptom of food intolerance that people tend to become addicted to certain food types. A good example is fridge raiding. If we look at the foods that are commonly eaten from the fridge they include milk, yoghourt, yoghourt drinks, cheese, chocolate, cheesecake, etc. These are all foods with a dairy content. There are very few who raid the fridge for broccoli.

Normally, food intake is regulated by feedback control mechanisms. Our need to eat diminishes as we take on board the normal food load. Addictive eating follows when the normal feedback control systems fail, or when we consciously or subconsciously override the feedback control because the craving cycle drives us to do this. It is sometimes explained away as comfort eating. The latter term seems to be have been derived from the creative writing seen in popular magazines. It has been adopted widely because it provides an excuse for food addiction.

Most people who eat too much know that they are eating more than the average person does. They also recognise that they are overweight and need to eat less, but find that they are compelled to continue eating by a restlessness that ceases only when they eat certain foods. It is suggested that it is a reduction in circulating cortisone that causes this restlessness and that eating the favourite snack triggers a rapid, though temporary, rise in circulating cortisone and brings back a mood of euphoria – the real reward for snacking. This is the addictive cycle. There may also be insulin-related effects that continue to turn calories into body fat, whilst reducing blood sugar levels and creating the need to eat more.

Cortisone is produced to protect against inflammatory responses to allergenic foods. The cortisone pulse provides a temporary feeling of euphoria. It is a key indicator of food intolerance reactions, and is thereby responsible for food intolerance sufferers repeatedly eating the foods that do them most harm. There is also a possibility, in some individuals, that a 'metabolite adsorption/depletion' mechanism reduces or inhibits the production of satiety hormones or neurotransmitters.

It might be productive to explore the causes of obesity rather than to attempt the development of anti-obesity drugs. Such drugs may reduce obesity, but they will not tackle the underlying disease processes that cause people to become obese. The exercise of willpower to reduce obesity becomes a very cruel sport, as long as the underlying causes of addictive eating remain unresolved.

This is where so-called detox methods play their role. In most detox diets, the principal allergenic foods have been removed and the bulk of the diet often consists of rarely eaten fruits or vegetables. By breaking the addictive cycle in a 'cold turkey' process, the diet helps the user to focus on the reorganisation of their approach to eating.

Addiction is generally accompanied by changes to thought processes that make it impossible for the sufferer to consider exiting from the addictive cycle. The very strangeness of many detox diets seems to help overcome this barrier to self-help. This may seem counterintuitive, but it does work for many overweight detox users. The problem remains that many users of detox diets then return to their 'normal' diet and are immediately caught in the addictive cycle once again. Detox diets have thereby come to have a poor reputation with many users, who end up regaining – or exceeding – their initial weight very quickly, after the dieting period.

Social consequences

Drug dependency
Where a food dependency cycle is established, there may be a stronger likelihood that the individual is also more susceptible to drug abuse, be it alcohol, tobacco, or something else. Using a drug is much like eating a foodstuff to quell the feelings of unease that follow falling blood cortisone levels between meals.

When young adults leave home, there is the possibility that, without a parent to fill the fridge with choice items and to prepare regular meals, they may experience much more prolonged withdrawal effects from not eating the foods that provoke their immune system reaction. They may then be more easily persuaded into drug abuse. The withdrawal of foods that are part of a masked allergy can produce unpleasant symptoms for several days, and this allows adequate time for experimenting with different, alternative substances.

Learning disorders
When contemplating the effects of atopic illness, it can be seen to have a huge and largely unmeasured and unmeasurable impact on the lives of a substantial percentage of the general population. Apart from the major physical impact of tiredness and/or muscle or joint pain, there is the impact of somatopsychic disorders induced by atopy.

So-called learning disorders often come with some of the other baggage of atopic disorders and should be viewed as part of the total package in terms of impact. So, when we add the impact of behavioural issues to other factors, a child will be seen to be severely handicapped in terms of educational competence within a normal setting.

The impact on a child is very limiting and, though the child may have high intelligence, they become imprisoned within the walls of the learning disorder. This is quite unlike the situation of a child whose learning is inhibited by the deficiency of their intelligence and the cruelty of this predicament can only be imagined. Additionally, the perceptual perspectives of the teacher and the child may be so different that the child's description of his or her situation may seem entirely weird.

Terms such as ADHD and Asperger's syndrome tend to be decided upon by psychologists who have little or no understanding of the root cause of the disorders. Help is then programmed from the same psychological perspective instead of being directed to the dietary cause. Dietary components, selected compulsively by the child, continue to damage the brain while this 'help' is instituted, and the use of drug treatment further obscures the link between symptoms and provocative foodstuffs. It is probable that most strongly autistic children are locked in this catch-22 situation.

Dyslexia

Dyslexia results from many different causes. It can be caused by physical impairment or by neurological impairment. Having been affected by both, the author has always struggled to make the best of a bad job.

The interpretive deficiency often appears to be inherited to some degree, but there are other possibilities. Hypernutrition, which is associated with bottle-feeding, is known to cause abnormally rapid body growth. It is not beyond the bounds of possibility that it also causes a more rapid growth of the components of the central nervous system. The growth rate of the brain must be tempered by subtle control mechanisms to achieve good function. If growth is accelerated, but enhanced organisational controls are not in place, then brain growth will not necessarily produce good brain function. This may affect the more complex functions of the brain differentially. The impact of immune complexes on brain development may affect a great many brain functions apart from the ability to read. It should also be borne in mind that eyes develop from outgrowths of the brain in the early embryo, and proceed to incorporate the dermal and mesodermal entities that they have induced to transform into the lens, the cornea, the sclerotic membrane, and the musculature of the eye.

Dyslexia is commonly seen as a simple interpretive issue relating to the recognition of written symbols, but is likely to be far more complex than most of us can easily understand.

Hyperactivity and hyperactive attention deficiency disorder (HADD)

There appears to be a knowledge gap on the status of conditions such as hyperactivity as a child develops into an adult. Hyperactivity seems to reflect the dissociation between the drive for action and the feedback and control mechanisms relating to the activity. A hyperactive child can be wildly violent without understanding what is driving it to such actions. When the state of hyperactivity ends there is still a lack of understanding of the compulsions during hyperactivity. The cause of hyperactivity is thought to be food-related, whether in relation to milk, cola, or candy, or perhaps such ingredients as tartrazine (E102). In these circumstances, the onset of hyperactivity is clearly an example of somatopsychic change.

In adult life, it might be presumed that similar mechanisms apply: that a similar dissociation results in normally pleasant people acting in aggressive, wild, and life-threatening ways. This is perhaps typified by road rage.

The consequences of hyperactivity are mostly felt by the affected family and, principally, the sufferer. Hyperactive behaviour is frequently punished as if it were wilful, when the opposite is most likely true. The hyperactive child feels punished for behaviour beyond his or her control, and feels confused regarding the forces that switch him or her from being 'normal' to being hyperactive. There is no amnesia or loss of perception about the hyperactive phase. Many parents unknowingly provide access to the psychotropic foods or additives by selecting meals for their children. Thus, the victim becomes the punished.

Diet and delinquency

When considering the hyperactive state, it does not take a giant cognitive leap to reach the conclusion that diet may be a major factor in delinquency. Within the protection of the home environment hyperactive behaviour is usually safe, but in the wider community it can lead to dangerous consequences and the involvement of the police and other authorities.

Therefore, we really must make the connection between eating particular foods, or a snack containing artificial colours and/or preservatives, and the risk of imprisonment for unlawful behaviour. It may be that stark. Obviously, the law does not easily consider such explanations. It may open the floodgates to potential false, but successful, legal defence for non-sufferers, with the defendant simply having to cite that he or she consumed a certain snack bar or piece of candy earlier in the day. Some unwitting 'expert witness' would then be called upon to define the region of hyperactive disorder into which the defendant fitted.

Rage and road rage

Considering what has preceded this section, it will be no surprise that the author connects road rage with diet. Of course, there are other causes of uncontrolled violent rage. The use of anabolic steroids to promote muscle bulk and definition is one of them. Nevertheless, we cannot ignore the probability that road rage could often be an adult version of hyperactive behaviour, or some other diet-stoked aberration.

So how should society deal with such a situation? Banning food colours and other chemical additives would only provide some protection, as most of the hyperactivity-inducing products are simple, natural foodstuffs for many people.

In many instances of diet-induced psychological change, the resultant behaviour can be said to be dissociative, or defective in integration. In the typical hyperactive situation, the child lacks the ability to integrate social responsibility into its behaviour, even though at other times it is clearly capable of such integration and easily manages what can be described as normal behaviour. In the case of road rage, the sufferer again fails to integrate social responsibility norms to guide control of the temporary urge to assault other road users.

Relationships

The impact of atopic disease symptoms extends beyond the individual sufferer. Families become affected, along with all those trying to engage in relationships with the affected person, whether as a child or as an adult.

The unlinking of action and control mechanisms in the mind can lead to conflicts with the community and law enforcement agencies. OCD tendencies can have a good and a bad impact on the individual and on those around them. What is obvious is that the associated issues of food intolerance reactions are unbelievably extensive and can have disproportionate impact on everyone.

I must be careful here or I will venture into the field of explaining the bulk of antisocial and criminal behaviour in terms of food reactions, and give an excuse to the majority of people held behind bars. I do not want to do that.

New scenarios

There are balances to the depressing picture given so far. Atopic children are often extremely intelligent and gifted in various ways. They often possess intuition that is unexpected in such young children. This might result from an early autistic barrier to interpersonal communication, leading to self-reliance for understanding the childhood environment. Dyslexia, for example, leads to an interpretation of the world that ignores the written word and allows the dyslexic person to develop an intuitive approach to interpreting their environment. Sir Norman Foster has made use of dyslexia in his design teams. A dyslexic architect provides a team with an intuitive approach to architectural design and an understanding of how a building works as an entirety, without the need to read labels and measurements.

A new society

As an ever-greater percentage of the UK population becomes affected by food intolerance reactions, precipitated by the widespread bottle-feeding of the newly born, and the capacity for atopic disease to be passed directly from mother to child through transplacental transfer, the population is gradually being transformed to a less and less healthy state.

Increases in the numbers of people suffering from HADD and autism make the education of the young more difficult, more expensive, and less effective. In old age, those with atopy consume ever more pharmaceuticals to offset the varied related symptoms. While the medical professionals remain in denial about their role in this emerging catastrophe, it is unlikely that any country will be able to develop policies that fully offset the problems of atopy within their populations.

A first step towards a new phase of health might be to restrict a small group of foods solely for infant feeding and to ensure that these are no longer components of the adult diet. This might be a far more effective way to improve the general health of the population than any number of pharmaceuticals.

Currently this would amount to a near extinguishing of the dairy and beef industry.

Unless a national programme is developed that includes the resources to consider all immunological impacts, it is unlikely that the general health of the population of the industrialised world will do anything but decline in terms of educational achievement and global competitiveness. Our residual advantage in a global economy will be to hold the knowledge of the future global impact of atopic illness.

It is not all gloom and doom, however. There is some positive impact from atopy, which, though difficult to evaluate, is definitely positive. One such example is the often-greater creativity of those with atopy. When brain function is slightly impaired, thoughts that have no apparent linear connection are often combined. It is from these new combinations of concepts that original ideas emerge. As a society, we are then faced with the dilemma of assessing whether or not a new combination of concepts is evidence of genius or of illness.

A brain unaffected by atopy tends to operate in a more linear fashion, and there is, consequently, a subtle difference in personality type between atopy-affected and atopy-free individuals. There is the possibility that we could progress faster in our development, as a civilisation, through the combination of contributions from those with atopy and from those without atopy working together. In a future, where atopy will predominate, there is no indicator for how things might turn out.

Research into diet-related disease outlined

The chemistry approach

The history of dietetics is dotted with a series of fundamental mistakes and oversights. The issue of the nutritional insufficiency of human milk has already been touched on. By any measure, this was a crazy piece of scientific deduction. The evaluation of the safety of formula milk products has been based largely on nutritional and biochemical (but not immunological) considerations. Any adverse reactions have been evaluated in terms of the infants consuming the products, with little consideration of any effects on later well-being. Infants with an eczematous reaction are still believed to 'grow out of it'.

The failure to recognise the significance of the substantial mass of immune system tissue associated with the alimentary tract, whether these are tonsils, Peyer's patches, or other lymphoid tissue masses, is another significant blunder. Early nutritional studies were negligent of the function of the immune system, while structures like the human appendix were generally considered to be troublesome remnants from a distant herbivorous antecedent.

As I mentioned earlier, vets have known for many years of the significance of colostrum in the survival of the neonates of a wide range of mammal species. It might, therefore, be thought that their very clear experience of this would indicate some pressing areas for investigation in newborn human infants.

The human child receives its most significant supply of maternal antibodies before birth, across the placenta, but maternal milk is still the only safe food in terms of the integrity of the child's gut and its immune responses. Maternal milk continues to be treated as self by the child's immune system. During gestation the child's immune system is adapted to the maternal environment, to prevent any immune attack or rejection reaction between the foetus and the mother. And the mother's immune system is similarly adapted during gestation to avoid any rejection of the foetus.

Modern research has tended to focus on component detail, with the repeated failure to reintegrate the model of the entirety of the life forms studied. Dietary studies have tended to look at the chemical composition of food without regard to immunological factors. On occasion there has also been a failure to recognise the significance of enantiomeric (mirror image isomeric) forms of food molecules, a recent example being studies of the efficacy or toxicity of vitamin E. It is relatively easy to manufacture mixed enantiomers of vitamin E, but only living organisms can succeed in producing the balance of enantiomers that are the most appropriate for nutrition and well-being. In excess, the manufactured vitamin E enantiomer mix is slightly toxic. The results of nutritional trials based on the use of this manufactured enantiomer mix (in excess) have been used to persuade the UK government to change some of the rules concerning natural food supplements. This is an example of bad science being used to facilitate bad regulations and to increase control by the medical industry and pharmaceutical companies.

Genetics

Early optimism about genetic research is only now being tempered by the experience of implementing therapies based on the genetic modification of human cells. A realisation is also beginning to dawn that genes are only a part of very complex cellular mechanisms and not the be-all and end-all of biological processes. The study of epigenetics is now pointing to how the body's cells control genes.

When a gene is found to be associated with a disease type like, say, autism, it is all too easy to fall into the trap of thinking that the gene is the one causal factor in the disease and then to describe the disease as 'genetic'. This has been the case with late-onset diabetes, which is now often labelled a 'genetic disease'. This avoids having to think about how the trend for diabetes has been shaped by particular environments.

In lands prone to repeated, prolonged drought and famine, where populations are repeatedly eliminated from vast land areas – as has happened many times in parts of India, for example, – natural selection for the diabetic tendency occurs because it enables the surviving individuals to subsist better on meagre rations. It is only when food returns in abundance that the diabetic tendency becomes a liability. So, perhaps 'environmental adaptation', rather than 'genetic disease', would be a better label to apply to late-onset diabetes. When damage to the regulation of the appetite caused by dairy intolerance combines with a genetic adaptation to an environment of food scarcity, there can be only one outcome in a world of food abundance: late-onset diabetes. A perfect storm, indeed.

If we think of genetic discoveries in a broader perspective, this may help to enlighten us rather than mislead us. Food intolerance is already a woefully complex field of study, as is atopic illness. And though there is little doubt that disease expressions will be paralleled by genetic patterns, there is not necessarily much benefit to be gained from that further detail of genetic information in the initial interpretation of symptoms and disease paths.

The language of immunology

One of the greatest handicaps in resolving immunological issues relating to foodstuffs is in the language of immunology itself. For example, we could say that, where there is an immunological food reaction, the subject has become immunised in some way against that food. Having an 'immune reaction' to a food ought to be the same as 'being immune' to the food, but in saying the latter we are, in a sense, contradicting the former. In terms of disease, both forms of expression would make sense. Being immune to a disease is seen as being able to mount a competent immune reaction. Being immune to a food may be viewed as being unaffected by some of its more indigestible components. The difficulties in how to express this sort of concept leads to difficulties in logical thought and the failure to identify real issues.

When we talk of 'passive transfer of immunity' in relation to disease, we are mindful of a protective function of the transferred antibodies carried either transplacentally or within the colostrum. It takes a twist of semantic logic to see that this transfer of antibodies can also cause food intolerance reactions in the recipient, if the mother is a producer of antibodies to foodstuffs. 'Passive transfer of immunity' sounds so much a gentle and benign process that may be protective to the recipient. It does not sound at all like the kind of process that might cause chronic illness in offspring, although this is exactly what it can do, generation after generation, through the maternal line. Wherever cross-feeding of female, newborn offspring has been practised, the transplacental transfer of immunity will perpetuate the damage through subsequent generations.

The passive transfer of immunity may also be followed by greater activity than the words suggest. Antibodies passed to the recipient may lead to further programming of the child's immune system, which then proceeds to manufacture its own version of antibodies to the same antigen. There are likely to be slight structural differences between the transferred antibodies and the recipient's own antibodies, due to genetic variation. Immune complexes formed by transferred antibodies are therefore likely to be different structurally from those formed from recipient-produced antibodies, and to have potentially different effects on health.

The term 'autoimmune' also requires a degree of semantic analysis, since it is often used to describe processes that are not strictly attacks by the immune system on the body. More often, the immune system is corrupted by environmental factors that cause a degree of self-damage in certain circumstances, but the damage occurs because of the interaction of immune system factors and environmental factors. If you remove exposure to the environmental factors the disease process often halts and, sometimes, repair is initiated. My own rheumatoid arthritis would be described as an autoimmune disease by some, but when I stopped eating foods containing cereals and sugar the arthritis also stopped and my joints began to repair. My hands show no indication of the arthritic state thirty-five years on, though my fingers were once swollen at the joints and twisted by the swelling. I can restart the arthritis at any time by again eating the implicated foods, but choose not to.

The above examples are presented to give an indication of the semantic pitfalls that await the analyst. They are in no way presented as a complete catalogue of the pitfalls of the language of immunology when applied to atopic illnesses.

Special interest groups

There has undoubtedly been some collaboration between parts of the UK dairy industry and the government in terms of looking at the synergies between dairy production and the promotion of healthy eating. For centuries, dairy produce has provided very beneficial nutrition in terms of fats, proteins, minerals, and vitamins. When cattle were fed on fresh grass supplemented by little more than hay, root crops, and salt licks, the fats in milk were very beneficial to human health.

The conversion of a pastoral activity into an industry has damaged our health. Cattle began to be fed on more processed foodstuffs. Cattle cake has at times included waste fats from the food industry, with protein and mineral additives, initially from fishmeal – and later, with tragic consequences, bones of slaughtered cattle and sheep.

Because of increased production, the milk could be processed into some of its components: butterfat, dried milk powder, lactose, casein, and whey powder. There was also a constant supply of milk for processing into infant formula feed, where it was mixed with a range of other ingredients to provide a nutritionally adequate diet for newborn and older infants. The exact mixture evolved in response to what manufacturers recognised as adverse reactions, which they obscured by describing reformulation as 'optimisation'. The component products of milk processing then came to be integrated into an ever-wider range of manufactured foods: sugary yoghurts, biscuits, cakes, instant meals, confectionery, and sports foods, etc.

The dispersal of manufactured products makes elimination dieting a complex process. It is easy to avoid fresh milk, but to avoid all foods of bovine origin involves reading the small print of every packaged food and understanding the origin of, say, casein, gelatine, or lactose.

Once newborn babies have encountered dairy milk derivatives, before their gut has matured sufficiently to cope with them immunologically, these foods become dangerous to them for the rest of their lives. Our immune system becomes corrupted by this encounter and is then capable of destroying our enjoyment of fresh, outdoor air, our ability to walk without pain, our ability to reason, our ability to learn, and our ability to enjoy social encounters. Nevertheless, we are still being told by government-funded bodies that we should all include dairy foods for a healthy, balanced diet. That dairy products are wholly beneficial now holds true for only a small and diminishing percentage of the population.

The widespread bottle-feeding of our newborn infants has changed our population in a way that will be difficult to reverse. Moreover, infant formula contains ingredients derived from a wide range of processed foodstuffs. The chances are that, if your immune system was provoked into reacting to milk-derived ingredients, then it may also have reacted to components of the other food sources in infant formula, while the gut lining was inflamed. As a population, we seem still to be affected by the addition of crude peanut oil to some infant formula milk brands, which has resulted in extreme peanut allergies that have been passed on through the maternal line to subsequent generations.

The medical industry

In Britain, the GMC-controlled orthodoxy does much to limit the development and improvement of the medical understanding of atopic illnesses. Medical students qualify, essentially, by being able to repeat the GMC mantras and by displaying some degree of competence in the application of such mantras to the treatment of patients.

An example of the result of such training is demonstrated in the typical procedure to establish whether a patient is a 'grumbler' or truly suffers from coeliac disease. In this situation, many a doctor will deploy the slightly dangerous technique of using an endoscopic device with the means to cut and hold a sample of duodenal mucosal tissue – a biopsy. If the subsequent examination of prepared sections of the mucosal biopsy reveals damaged villi in the sample, then the patient is declared to be coeliac. How much simpler and safer it would be to place the patient on a gluten-free diet for a few days, and then to reintroduce gluten into the diet to determine the body's reaction. Then we would know that gluten was an issue and not some other foodstuff, or pathogen, that might also cause damage to the gut villi. As I write, it is still being described thus on a medical website: 'A small bowel biopsy is the gold standard for diagnosis of coeliac disease'.

The tobacco industry

Why mention the tobacco industry in a book on food intolerance? Well, mostly to demonstrate the power of industrial lobbyists on our politicians and government establishments, against the well-being of the nation. The story is like that of the asbestos industry.

Despite the clear and rather obvious damage that had been done to the health of smokers by tobacco smoke, the industry funded a great deal of 'scientific research' over which, as fund provider, it had a great deal of control. In the days when the tobacco industry was still denying any link between lung cancer and smoking I was still at school, and I remember a school chemistry teacher telling us about his earlier research into the properties of the components of tobacco tar, especially benzopyrene. 'This,' he affirmed, 'is a powerful carcinogen, and is breathed in with every puff from a cigarette and by all those around the smoker.' That was in 1964.

Such research results were published for all to read. In the nineteenth century it was benzopyrene that acted as a potent carcinogen that caused young chimney sweeps to develop cancer of the scrotum in unusual numbers, after sitting on their brushes to clean inside chimneys. It is an accepted product of the combustion of organic materials.

Our UK New Labour government finally ignored the lobbyists and Bernie Ecclestone (a Labour Party donor) and put smoking out of work and out of advertising in 2007. That is the power of industrial lobbying in one of the world's key democracies: more than forty years from knowledge to full action.

The ethical pharmaceutical industry

If the tobacco and asbestos industry could lobby successfully, just imagine the capability of the pharmaceutical industry. It is not even starting on the back foot. It is functioning for the benefit of all of us ... is it not? I mean, what harm could it do to anyone?

Well, the death toll from ethical pharmaceuticals worldwide is well above the hundreds of thousands. The increased suffering the industry causes is not even measured. Thalidomide was just a sample episode in the provision of needless poisons. Some of its limbless victims are still around to remind us of the supplier's carelessness and lobbying power. Its manufacturers are still doing very nicely because of AIDS and the continuing ravages of leprosy, for which thalidomide has now proved to be of some use.

Some statins are recent heirs to the tradition of marketing pharmaceuticals – in this case, to provide the UK government with statistics to demonstrate a 'health benefit' from reduced 'blood cholesterol' levels. Here we have the pharmaceutical companies and the government working hand in hand to give government ministers a perceived political advantage – and the drug manufacturers an increased profit.

This is a very cosy situation ... except that a significant percentage of statin users are known to be experiencing extreme muscle and joint pain, memory loss, and other symptoms. Maybe there is another marketing opportunity here. Targeted painkillers, perhaps: a poison to mask the effects of a poison, and not for the first time. My dislike of statins is linked to the use of pharmaceuticals to mask atopic disease symptoms, while the underlying disease process continues.

Research funding bodies

Funding bodies should be seen to distribute funds in a sensible way. Generally, this means following the establishment view of the world. This immediately reduces the potential scope for research. There are few organisations offering funding for atypical research, but even those that do exist are wary of fund-seekers who have already received rejection elsewhere. Just now, the theory is that if you put 'global warming' or 'climate change' somewhere prominent in your funding request, you stand a much better chance of receiving those funds. The key point is that funding is as much a victim of fashion, as so much else is today. It is not so much that our funding organisations examine the full scientific context and validity of research, as that they look for the words that are currently in fashion. It is as if the western world is steadily unlearning the principles of science and replacing science with a destructive mixture of fashion, marketing, and belief.

The requirement for research papers to be appraised by peer review can, in its own way, be a source of corruption in the scientific process. At present, anyone daring to question the validity of the arguments used by the 'global warming through CO_2' lobby is immediately attacked and even driven out of post for offending the presumed consensus. This does not produce better science, as time will show.

The 'National Institute for Controlling Expenditure' (in medicine) is a common interpretation of the acronym NICE. The official version is about clinical excellence, but few practitioners would accept that as a true description. In the cosy relationship between government and pharmaceutical companies, NICE is about rationing new drugs on a statistical cost-benefit basis. This is a new way to undermine the clinical independence of doctors. Governments have now lost the pioneering zeal of those who initiated the welfare state and the health service, and are making decisions based on voter appeal and political advantage instead.

Self

We all interpret what we see, hear, and learn in terms of our own experience. If you present a smoker with the overwhelming evidence of the health hazards of smoking they will be very likely to counter your argument with a story about a person they know who smoked sixty a day, drank like a fish, ate chips every day – and lived to be ninety-five. Statistics mean nothing to an addict. Similarly, if you happen to work for the pharmaceutical industry or the dairy industry, you may well have filed this book in the paper-recycling bin many pages ago. If, on the other hand, you have suffered atopic symptoms of one sort or another and have rarely found medicines to provide good relief from your symptoms, you may be persuaded to read to the very last page in the hope of finding some useful truth.

In the light of their training and expertise, I am fascinated by the response of practising doctors and nutritionists. I am a problem in the GMC view of the world, as someone who had rheumatoid arthritis of sufficient intensity to be forced out of a career by it and to be certified disabled, but who became free of signs of this past illness without the use of corticosteroids or any other drugs. The usual conclusion is that the arthritis that I suffered was not typical. An alternative viewpoint is that I must be a hypochondriac. My viewpoint, which is that I became free of arthritis because I changed my diet, is of no interest to such practitioners in this context. The fact that I can induce the old symptoms by consuming the foods that I have eliminated from my diet (cereals, or indeed any of the grass family, from wheat and rice to sugar cane) adds a complexity outside the realms of their training that they would prefer not to consider.

I would have expected that a medical practitioner, on being told that I can lose or bring on rheumatoid arthritis symptoms by dietary changes, would be keen to check this out experimentally and scientifically. However, I have never been invited to take part in such an investigation. The same applies for my friend and former asthma sufferer, David. It appears that it might confound the normal GMC mantras, the sacred cows of modern medicine. And it might disrupt the standard practice of ethical pharmaceutical reselling.

In place of a willingness to investigate scientifically, different orthodox practitioners have offered me Librium, anti-arthritics, asthma treatments (I do not have asthma symptoms), desiccated liver tablets, blood pressure modulators, and sympathy. Alternative practitioners have done a little better, especially with the sympathy, but seem to be equally bound, by their training, from the ability to think more widely.

One practitioner, a medical herbalist, has been able to take the link between diet and symptoms as a starting point and then develop further strategies towards helping me improve my health. With this practitioner's help, I am now more in control of my life and health. An allergy clinic has helped resolve my tendency to episodes of extreme high blood pressure, and I may well owe my continued existence to their special skills.

With the continuing threat to the marketing of unlicensed herbal products I might soon have to risk purchasing directly from overseas, through the Internet. This is presented to us as progress towards safer medicine.

Industry lobbyists have turned logic on its head. This is par for the course for politicians, for as long as the pharmaceutical industry's inducements to them are sufficiently persuasive.

Resolution

Plan of action
How to respond to the current atopic disease epidemic will have to be the subject of another book.

Man, and medicine
Domestic pets are now suffering the same fate as their masters and are displaying atopic disease symptoms including eczema, arthritis, and heart disease. This is unexpected, to a degree, because vets have always appreciated the need for maternal colostrum followed by a further period of feeding on maternal milk for health and survival in most mammal species. Some breeders of pet dogs use substitute milks to keep up the production numbers of currently fashionable breeds. They sell the pet dogs as puppies, and so have little regard for the health of the creatures in later life. There has been some development of substitute feeds for some farm animals. Lambs have often been removed from the field to be hand-reared, either as pets, or because the ewe cannot or will not feed them.

Western man, on the other hand, has remained obdurate in respect of recognising his place in the world of living things. We knew little about other living apes in terms of infant and adult nutrition until such research as that undertaken by Jane Goodall, on chimpanzees. In fact, our hesitation to study the other living apes in the wild has certainly held back our understanding of our own natural biology. Jane Goodall observed that infant chimps first chewed and swallowed solid foods at about four months, and that final weaning was achieved by the end of the fifth year.

The study of human nutrition accelerated under the pressures of the two world wars, and was undertaken by biochemists. It is only in comparatively recent times that the significance of the immune system in dealing with foodstuffs has come to be recognised.

Biochemistry has brought us an understanding of the physiological components of living systems and has confected medicines, pesticides, and herbicides that were important during times of war, and subsequently in the control of epidemic disease. It provided details of the necessary components of diet in chemical and physical terms: amino acids, carbohydrates, fats, vitamins, minerals, and roughage. It failed, however, to incorporate the immunological implications of feeding newborn humans on processed foodstuffs rather than on mother's milk. This failure is reflected in the UK Department of Health, report 47, published in 1996: *Guidelines on the Nutritional Assessment of Infant Formulas*, in which there is little, if any, reference to the reactions of our immune systems to foods.

Today a great many medicines owe their profitability to the stream of symptoms that follow the bottle-feeding of infants. There are instant palliative treatments for colic, eczema, asthma, migraine, irritable bowel, arthritis, etc., but there is little understanding of the causes of or the relationships between such atopic disease symptoms.

In other cultures, notably in parts of Indo-China and South America, a science of medication has evolved that relies on mostly herbal sources. Each culture has had to determine the most effective remedies from among the available endemic (or introduced) species of plant and other natural resources.

We now know that chimpanzee tribes each evolve their own complex natural pharmacopoeias, though we are only beginning to understand their lives and customs. Such systems of natural and herbal therapy develop within a generally stable ecosystem over generations of experience, and become embedded in the culture of the particular tribe or civilisation. At most times during the last 30,000 years the most stable region of the Earth for the development of such therapeutic systems has been the tropics. Ice ages, volcanic eruptions, and other geological and cosmological events have made the so-called temperate regions far too variable in climate for such continuity of culture. So, we might expect the most developed natural pharmacopoeias to occur in more tropical regions, and that seems to be the case.

In temperate – now industrialised – regions, we started with a disadvantage when it came to developing natural pharmacopoeias. We therefore responded favourably to the opportunities that were eventually provided by the development of Western science. We now tend to look on natural treatments as inferior to those provided by science: less technically advanced, and less predictable. However, a well-developed natural system of healing has evolved in both Northern Europe and North America.

The discovery of the 5,300-year-old frozen remains of Ötzi the Iceman in an Alpine glacier in 1991 led to the further discovery that acupuncture therapy may have been used to treat his ailments and that tattoos on his body may have indicated the appropriate acupuncture points. An identical therapy is in use today by so-called alternative practitioners. This may make us question the application of the term 'alternative'. Perhaps it could, more reasonably, be said that modern medicine now uses a 'chemical alternative' therapy.

Healing vs ethical pharmaceutical reselling

So, now we have identified that much of today's medicine is of a 'chemical alternative' kind, what are the long-term consequences of employing it? Are we much healthier and more resilient now, or are we more prone to long-term illness and more susceptible to pandemic diseases?

One certain consequence is the empowerment of large pharmaceutical companies to usurp and control medical, pharmacological science. Nowadays, therapies that may have been used for more than 5,000 years are labelled as 'untested' by lobbyists for the pharmaceutical industry. They cannot define exactly what this means, nor give a reasonable account of why so many of their so-called tested and approved pharmaceuticals have had to be withdrawn, after death and debilitation has hit so many recipients of these chemically synthesised drugs.

This area of medicine and science is no longer driven by a desire to cure the sick, but a need to increase profit, yield, and market share in a global economy. It is interesting to note that in some cultures it was often the case that the physician or healer was paid to maintain the health of his clients and received no payment, or no increased payment, for treatment during periods of their ill health.

Asthma treatments and their marketing

The rejection of so many natural remedies that have been tested by time, rather than by modern science, could become a disaster for human health. So many plant-derived treatments are curative rather than palliative, and those that are palliative are commonly free of unwanted side effects.

Compare this with the approach of big business. Are profits best maintained by curative or by palliative treatments? Is it more profitable to rid people of, say, asthma, or to sell them repeat prescriptions of inhalers for the rest of their lives? Considering that much asthma could be eliminated by simple dietary change, it is depressing to watch the dependency and loyalty of such patients to the pharmaceuticals that keep them suffering. In the final analysis, the limitations of the medical profession in the science of healing are allowing pharmaceutical manufacturers to hold sway.

A general practitioner's standard appointment system gives the patient about seven minutes of the doctor's time – just about enough time to list the symptoms and scribble a prescription. This is just what the pharmaceutical companies would want. The GP has no time to consider family history, lifestyle choices, diet, addictions, housing, and family pressures – the factors that most affect our health.

On the other hand, if you visit an acupuncturist, a herbalist, or a natural therapist, you are likely to be subjected to an initial hour-long interview, which touches on all the areas listed above: family history, lifestyle choices, diet, etc. Your treatment is then followed by an in-depth analysis of its consequences.

So, the modern, 'chemical alternative' to treating disease, suffers from major faults:

- It is based on recent, commercially funded science, while ignoring much of the wisdom of the past.
- It tends to ignore the cause of atopic illnesses while focusing on the symptoms.
- Doctors have little time to do more than list symptoms and issue a palliative prescription.

The major effect is that symptoms are pharmaceutically masked, the underlying disease process continues unrecognised, and the patient stumbles from one set of discomforting symptoms to another, until the end of a painful old age brings peace. Pharmaceutical science, incorporating aspects of ancient herbal medicine, is becoming the legal, intellectual property of major pharmaceutical companies. The general community is gradually being deprived of its tradition of knowledge by the dogma of 'science-based' medicine and the endeavours of pharmaceutical companies, who are exploiting their niche in this process.

If we were to reappraise the NHS, we might consider the fact that much of the cost of maintaining the population in a state fit enough to attend work is linked to atopic disease symptoms. We might therefore reallocate some of the funds to supporting independent research and treatment of these diet-related disorders. Many children today are unwell because of malnutrition caused not by lack of food, but by an overabundance of junk foods. A huge programme of re-education is required, not only for parents and their children but also for the medical professionals.

Just as some schools are now trying to put junk foods out of reach of their pupils, so we should put free, palliative treatment beyond the reach of those whose symptoms can be relieved by dietary change, while improving palliative care where such alternative relief cannot be found. The only people to lose out from this would be the pharmaceutical and prosthetics manufacturers. Everyone else would win, in the short term and in the long term.

The ethical course of action is to heal, rather than to dose to suppress symptoms.

The big question is about why medical science has not arrived at this truth long ago. The answer is complex. Science is an art and a faith that may be well or poorly practised. It is not an absolute, however much we like to think it is.

Much science is accepted because it has the name of a distinguished scientist behind it. Recent child abuse cases in the UK have demonstrated that having a distinguished professor as an expert witness is not necessarily the path to truth and wisdom. However, there is no profession that is free from prejudices. Much research is now supported by commercial or political organisations having a mission statement that funded scientists must comply with. Science is constantly in a state of sorting itself out, often with new dogmas replacing old dogmas.

Science is about truth in knowledge, not just about methodology. Today's insistence on double-blind trials and peer-reviewed publication helps to limit science to the holders or the givers of funding for that complex process. I wonder how many double-blind, experimental, and statistically valid samples of falling apples Newton would be required to produce to describe his theory of gravity, if he were to be researching that today …

In the present intellectual climate, where blame has a higher score than scientific accuracy in any discussion, it is very much an uphill struggle to revisit old paradigms of medicine with a view to revising or overturning them. Even science itself has become so constrained in its function, by those who have a stake in controlling it, that it has become nearly impossible to present new thinking without huge amounts of quantitative investigation and statistical analysis. Perhaps we need some journals to enter more into the spirit of publishing 'ideas'.

I have attempted to put on show here what is in my heart and mind, based upon lifelong observation and deduction that has been supported by reading a broad range of research papers published in the relevant scientific journals. Traditional therapies have been poorly presented in the standard journals – largely, it would seem, because mainstream medicine has driven qualified physicians away from such practice. Some fine, deep-thinking practitioners have been driven to emigrate, such has been the pressure from the medical establishment that has been created by ignorance and fear of both ancient knowledge and emergent new paradigms.

Modern medicine should not overlook the huge number of patient deaths and injuries for which it has been responsible, in allegiance with the pharmaceutical industry. The huge number of people who continue to suffer atopic illnesses, because prescribing palliative pharmaceuticals is more profitable than finding a way to stop the disease process, can be added to the thousands and millions damaged by modern medicine.

A revised definition of atopic illness

Use of the terms 'atopy' and 'atopic disease' has varied over time, and has become somewhat vague. I suggest that it may be appropriate to define these terms as 'disease linked to the production of antibodies against commonly consumed foods and the symptoms that may be expressed as a result of the consequent reactions of the immune system, wherever they may be expressed in the body'. This revision sweeps most of our Western diseases under the atopy umbrella. I feel that this may help in planning an escape from the symptoms that afflict so many of us in the West and around the world.

Atopic disease begins with the feeding of newborn infants on confected milks. Antibodies are produced against these foods by the infant's immature immune system, and this production of antibodies continues throughout life for as long as such foods are consumed. Disease symptoms appear somewhere around the body, or asymptomatic damage occurs, whenever the offending food groups are consumed. I have assumed that it must be the antigen-antibody complexes having an individual structure that determines the type of symptoms that occur. For example, if these complexes adhere by chance to the tissue surfaces of the islets of Langerhans, then destruction of the islets follows and the production of insulin diminishes or ceases, causing a diabetic crisis - early-onset diabetes.

Thus, in many ways, diseases such as early-onset diabetes, or, for example, asthma, are symptoms of atopic illness rather than specific diseases in their own right. At the same time that these symptoms are displayed, there may be as yet symptomless damage occurring to the cardiovascular system or to the gut, the prostate, the breasts, or other parts. At a later stage, this other damage may be expressed as heart disease, and as cancer. And all of this arises from that single but disastrous change to the nutrition of our newborn children. This process is described from an epidemiological perspective by Maureen Minchin in her book, *Milk Matters: infant feeding and immune disorder*, published in 2015.

Undoing the damage

So far, I have described my observations and selected research findings relating to the effects of the bottle-feeding of a very wide sample of this world's infants.

I have explained how the health impacts of bottle-feeding manifest as a series of symptoms that tend to change during an individual's life – from colic to cardiovascular disease, for example. I have indicated that the food intolerances that are induced by proprietary, formula milks tend to relate to the component food families in the 'milk' formula. These food intolerances can be inherited through the maternal line by immune system programming from mother to child. Sometimes there is a simple inheritance of similar symptoms and sometimes there is a step change in the severity of the reaction from one generation to the next, such as is the case of an extreme reaction to peanuts or the extreme reaction to rubber latex (this was probably more common when the teats of feeding bottles were made of natural rubber).

The huge biological change in the nutrition of human infants in industrialised societies has gone largely unrecorded. What ingredients have been fed to infants in the place of maternal milk and the consequences of this change are largely obscured, although I think it may have been realised that the addition of peanut oil to some milk formulas is in some way related to acute peanut allergy, for example.

From Liebig's formula to modern commercial brands we have little recorded evidence of the lifetime effect on consumers of such infant foods, except where that life has been dramatically foreshortened. On the other hand, we are in a situation where progressively more children are diagnosed as asthmatic, or autistic, plagued by glue ear, or simply obese. As a society we are beginning to wonder why. All kinds of suggestions are put forward, often because of some statistical correlation – traffic fumes and asthma, in Japan, for example. This book is an attempt to show that there is an overarching relationship between infant nutrition and atopic disease symptoms.

We must apply our logical reasoning beyond the simplistic 'dust mites cause asthma' style of approach to one where we say, 'Dust mites have been in our houses ever since we had houses,' 'Only now are we linking dust mites to asthma,' 'Why is it that in the same house one sibling, even an identical twin, suffers from asthma while the other appears clear of symptoms?' and, 'Why does asthma become more of a problem after the sufferer has eaten dairy-rich foods or beef?'

Only when we get into such detailed examination will we be able to begin to see the true *cause* of asthma, rather than identifying the *trigger* of dust mites as an 'obvious' cause. Dust mites may trigger an asthma attack, but they do not cause the underlying disease of asthma and they are only one of many potential triggers of asthma attacks.

Modern medicine tries to appeal to the populist view. It uses simplistic logic, where this helps to get a message across, especially when the message is, 'Use your inhalers. We are here to solve life's little problems for you.' This is how modern chemical alternative medicine becomes a pernicious limitation to our knowledge and understanding of diseases and healing. It also indirectly supports a huge market in peripheral products such as dust mite sprays, air ionisers, air filters, ventilation systems, nebulisers, humidifiers, dehumidifiers, inhalants, and so on, in the case of asthma, none of which would seem necessary if the sufferer simply avoided foods of bovine origin.

How did we become so obsessed with cow's milk and all its derivatives? And why did we ever think it would be good for newborn children? When will our medical gurus declare that it might be a good idea to avoid dairy products and beef products generally? We are so aware of the dangers of the early weaning of neonates of most other species of mammal. We have tended to take better care of kittens and puppies than of our own infants in this regard. However, where such pets have been weaned too early onto various cow's milk formulas, they have gone on to display very similar atopic symptoms to our own, including eczema, arthritis, and cardiovascular disease.

We do not need beef or dairy products. Major civilisations around the world have flourished without them. Diseases such as breast cancer are rare in many populations that have no regular access to dairy products or feeding bottles. Neither do we need dairy milk substitutes with added calcium and vitamins. Calcium is not usually in short supply, except perhaps in some manufactured foods.

How can we correct the current situation? Even if we go back to earlier, traditional ways of child-rearing and dietary choice, we may not fully correct our misdirected immune systems. There are increasing pressures to add probiotics and other substances to formula milk in order to correct a tendency to atopy that has reached epidemic proportions. However, the more we add to feeds for newly born and very young infants, the more we are likely to extend the range of foods that cause illness and the variety and the abundance of atopic symptoms.

The key is to not provoke the infant immune system with items that may become part of a future diet. Mother's milk will always be the kindest resource we have for newborn infants – the one safe food. A mother's breast milk meets the criteria for non-provocation and immune integrity, and supplies the essential nutrients and water. It was designed for the job by evolution, and we cannot design a better product. Nor can we design one that meets all the criteria for the health of the infant immediately after birth and throughout its life.

Liebig's formula was a disaster for many infants and, of those who survived, the child-bearers managed to pass on their resultant corrupted antibody response through succeeding generations. Instead of trying to improve on a disaster, we should have understood that all alternatives to breast milk might cause harm. Moreover, it would be a harm that can be passed from one generation to the next. There is no sensible and safe replacement for breast milk to feed a newborn child. There are only potentially harmful alternatives. There is only one good option, and that is for a mother to feed her infant on human milk – preferably her own.

The development of farming and the domestication of cattle in the Fertile Crescent of the Middle East provided humankind with vast new food resources to exploit. By mistakenly giving some of these foods to our newborn infants in recent generations, we have created a world in which many of them will no longer be able to benefit from this bounty. If we continue the present Western systems of infant nutrition we will have not only a much greater burden of atopic disease symptoms within the population but also the beginnings of a food supply problem for those who can no longer consume common foods without suffering disabling atopic symptoms. Some will see this as the biggest new opportunity for major pharmaceutical companies worldwide. Western medicine and state health systems have helped introduce Western diseases to many of the peoples inhabiting this planet. In time, the cost burden of palliative care for atopic disease sufferers has the potential to bankrupt healthcare systems around the world.

If you have found this book interesting, come and join the conversation at:
https://thekindnessofhumanmilk.com/

Summary statements:

- Western diseases are essentially atopic diseases, though most are not currently described in that way.
- Atopic diseases are caused by feeding any number of foods other than maternal human milk to newborn babies.
- Atopic disease is passed from generation to generation down the maternal line as an acquired characteristic, or characteristics, through transplacental transfer.
- A mother with atopy will always pass the potential for atopic disease to her children.
- The symptoms of atopic disease depend on maternal health, parental genetic variability, and chance.
- Asthma, eczema, glue ear, gluten intolerance, rhinitis, rheumatoid arthritis, osteoporosis, lower bowel, breast, and prostate cancer, early-onset diabetes, autism, depression, obesity, atheroma, heart disease, and many other diseases are consequences of atopy linked to bottle-feeding, or to having an atopic, biological mother.
- A father with atopic disease may pass the genetic potential for specific symptoms, but not atopy itself, to his children.
- There is no known, safe cure for atopic disease within Western medicine. Only the suppression of certain symptoms is possible. The avoidance of provocative foods may lead to increased immune sensitivity at first, due to reduced cortisone levels, but may halt or slow the progress of some atopic symptoms.
- Asthma normally responds to a diet free of foods of bovine origin. It is likely to return powerfully if foods of bovine origin are again consumed after an interval.

Glossary

This glossary defines terms in the way that they are used in this book.

Adsorption
The process of becoming bound to a surface.

Amyloid protein
A basic, structural protein that has the ability to aggregate into fibrils.

Antibody
A complex molecular structure produced by immune system cells called B cells, after provocation by an antigen. It is capable of specifically binding with that antigen as a first step towards the destruction and elimination of that antigen.

Antigen
A molecular species that is capable of inducing the production of antibodies in the host. In a neonate's gut, any food other than mother's milk is likely to be treated as an invasive antigen. In time, the infant gut matures to deal with a mixture of food sources without alerting the body's immune system and weaning can be done safely, but any previous antigen exposure is likely to produce a persistent antibody production response well into adulthood.

Antigen-antibody complex
An immune complex formed from the integral binding of an antibody to an antigen or antigens. The action of immune complexes varies from individual to individual, perhaps genetically. It appears that immune complexes can have structures that make them bind to specific tissue components in such places as articulating joints, bronchiolar structures, the islets of Langerhans, or colder regions of the body such as the skin. The conjunction of antigen and antibody normally triggers the complement reaction process.

Atopy
In the original definition of this term there is the principle that symptoms are 'out of place' or 'without place'. In modern interpretation, there is a fashionable link to genetic predisposition. I have considered the idea of 'lacking a place' to refer to the link between cause and symptom, and I use the term 'atopic' to describe disease symptoms that have arisen through the corruption of the immune system by environmental or nutritional influences. Thus, early-onset diabetes may be considered an atopic symptom of an immune reaction to dairy foods and beef.

Autologous transfer
A transplant of cells or tissue from one part of a body to another part of the same individual.

Basal lamina
The layer of a basement membrane directly in contact with epithelial cells.

Basement membrane
Epithelial cells are anchored in a basement membrane consisting largely of structural protein fibrils embedded in a matrix of glycoproteins. This membrane separates the surface epithelium from most underlying cellular structures but allows diffusion of oxygen and nutrients from nearby blood capillaries.

Cavies
Another word for guinea pigs. Cavies are rodents that are native to South America.

Complement reaction or cascade
When antibodies become attached to their specific antigen the antigen-antibody complex that is formed serves as a focus for a series of reactions designed to immobilise and destroy an invading organism. This is the complement reaction/cascade. When the reaction is in response to ingested antigens, the response may give rise to the destructive inflammation of gut

tissue and local tissue. This pattern of complement reactions forms the basis of many allergic symptoms and type I diabetes. The localisation of inflammation may result from the antigen-antibody complex being selectively adherent to particular tissue components.

Dermis
The layer of skin tissue beneath the epidermis.

Enantiomers
Enantiomers are optical isomers (see below) or mirror image copies of a molecule of a chemical compound.

Enterovirus
A type of virus transmitted through oral, nasal, or faecal contact. Polio is an enterovirus.

Epigenetics, epigenes
Epigenes control the expression of genes. Epigenetics refers to external modifications to DNA that cause genes to be active or inactive. These modifications do not change the DNA sequence, however.

Foods of bovine origin
Foods deriving from dairy products or beef, including but not limited to cow's milk and its products, steak, gelatine, dripping, casein, whey, Bovril ™, etc.

Haematopoietic
Relating to the production of blood cells.

Ischaemia
The restriction of blood supply.

Isomer
An isomer is a compound with the same molecular formula as another molecule, but with a different molecular structure.

Lumen
An enclosed space, such as that enclosed by a tube or the gut wall.

Microchimeric
Including a small percentage of cells from another individual.

Microchimerism may be due to the transplacental transfer of cells between mother and foetus or between twins.

Narcoleptic

Narcolepsy is a neurological condition most characterised by a tendency to fall asleep involuntarily during the daytime.

Proteolytic

A protein-digesting substance, usually an enzyme.

Psychosomatic

Bodily symptoms caused by mental stress.

Somatic

Relating to the body, as distinct from the mind.

Somatopsychic

Mental symptoms caused by bodily illness.

Transplacental transfer

Carried across the placenta to or from the developing foetus.

Ungulate

A hoofed mammal.

Villi

Simple, finger like projections of the endoderm of the small intestine that extend about 1 mm into the gut lumen, including blood capillaries, to assist the absorption of digested foods and lipids by increasing the surface area of the gut. Villus-type structures also exist at the interface of the vascular systems in the placenta.

Bibliography – Books

Brostoff, Jonathan, and Challacombe, Stephen J., 2002, *Food Allergy and Intolerance.*
Freed, David L.J. (Editor), 1984, *Health Hazards of Milk*
HMSO, Department of Health Report 47, 1996, *Guidelines on the Nutritional Assessment of Infant Formulas.*
Institute of Medicine of the National Academies, *Infant Formula – Evaluating the Safety of New Ingredients*, Washington DC, National Academies Press.
Le Breton, Marilyn, 2001, *Diet Intervention and Autism.*
Mackarness, Richard, 1958, *Eat Fat and Grow Slim.*
Mackarness, Richard, 1976, *Not All in the Mind.*
Mackarness, Richard, 1980, *Chemical Victims.*
Minchin, Maureen, 1992, *Food for Thought: A Parent's Guide to Food Intolerance.*
Minchin, Maureen, 2015, *Milk Matters: Infant feeding & immune disorder.*
Minchin, Maureen, 2016, *Infant Formula and Modern Epidemics: The milk hypothesis.*
Mumby, Keith, 1985, *The Food Allergy Plan.*
Pollard, Tessa M., 2008, *Western Diseases – An Evolutionary Perspective*, Cambridge University Press.
Randolph, Theron G. and Moss, Ralph W., 1981, *Allergies: Your Hidden Enemy.*
Young, Charlotte, 2016, *Why Breastfeeding Matters.*

Bibliography - Papers

Gerstein, H.C., (1994), *Cow's milk exposure and type I diabetes mellitus. A critical overview of the clinical literature*, Diabetes Care, **17**(1), 13–9.

Verge C.F., et al., (1994), *Environmental factors in childhood IDDM. A population-based, case-control study*, Diabetes Care, **17**(12), 1381–9.

Maródi, L., (2006), *Neonatal Innate Immunity to Infectious Agents – Mini review*, Infection and Immunity, **74**, 1999–2006.

Stevens, E. E., Patrick, T. E., & Pickler, R., (2009), *A History of Infant Feeding,* The Journal of Perinatal Education, **18**(2), 32–39.

Shoenfeld, Y., et al., (2005), *Accelerated Atherosclerosis in Autoimmune Rheumatic Diseases*, Circulation, **112**, 3337–3347.

Hendaus, M.A., Jomha, F.A., Ehlayel, M, (2016), *Allergic diseases among children: nutritional prevention and intervention,* Therapeutics and Clinical Risk Management, **12**, 361–372.

Torre, Jack C. de la, (2004), *Is Alzheimer's disease a neurodegenerative or a vascular disorder? Data, dogma, and dialectics*, The Lancet Neurology, **3**, 184–190.

Wakefield, A., (2005), *The Seat of the Soul; The Origins of the Autism Epidemic*, A talk presented at Carnegie Mellon University, 17 November 2005.

Lanou, A.J., (2009), *Should dairy be recommended as part of a healthy vegetarian diet? Counterpoint*, Am. J. Clin. Nutr., **89** (suppl.), 1638S–42S.

Anderson, H.R., Gupta, R., Strachan, D.P. and Limb, E.S., (2007), *50 years of asthma: UK trends from 1955 to 2004*, Thorax, **62**, 85–90.

Cañas, C.A. et al., (2016), *Is Bariatric Surgery a Trigger Factor for Systemic Autoimmune Diseases?* Journal of Clinical Rheumatology, **22**, (2), 89–91.
Yan, P. et al., (2011), *Biological Characteristics of Foam Cell Formation in Smooth Muscle Cells Derived from Bone Marrow Stem Cells*, Int. J. Biol. Sci., **7**, 937–946.
Cardwell, C. R., Stene, L. C., Ludvigsson, J., et al., (2012), *Breastfeeding and Childhood-Onset Type 1 Diabetes*, Diabetes Care, **35**, 2215–25.
Tyndall, A.J., Joly, F., Carbonne, B., et al., (2008), *Pregnancy and childbirth after treatment with autologous hematopoietic stem cell transplantation for severe systemic sclerosis requiring parenteral nutrition. Ethical issues*, Clinical and Experimental Rheumatology, **26**, 1122–1124.
Chapman, D.P., Perry, G.S., Strine, T.W., *The vital link between chronic disease and depressive disorders*, (2005), Prev. Chronic Dis. [serial online].
Kolev, M.V. et al., (2009), *Implication of Complement System and its Regulators in Alzheimer's Disease*, Curr. Neuropharmacol., 7(1), 1–8.
Prentice, A, *Constituents of human milk*, United Nations University.
Stuart, F.A., Segal, T.Y., Keady, S., (2005), *Adverse psychological effects of corticosteroids in children and adolescents*, Arch Dis Child., **90**, 500–506.
Piras, I.S. et al., (2014), *Anti-brain antibodies are associated with more severe cognitive and behavioral profiles in Italian children with Autism Spectrum Disorder*, Brain Behav. Immun., **38**, 91–99.
Tucker, L.A., , (2015), *Dairy Consumption and Insulin Resistance: The Role of Body Fat, Physical Activity, and Energy Intake*, Journal of Diabetes Research, Article ID 206959, 11 pages.

Health and Wellbeing Division Department of Health, (2013), *Consultation: Draft Statutory Instrument – The Infant Formula and Follow-on Formula (England) (Amendment) Regulations 2014,* HMSO.

Inoue et al., (2012), *Infant feeding practices and breastfeeding duration in Japan: A review*, International Breastfeeding Journal, **7**, 15.

Lasso-Pirot, A., Delgado-Villalta, S., Spanier, A.J., (2015), *Early childhood wheezers: identifying asthma in later life*, Journal of Asthma and Allergy, **8**, 63–73.

Bégin, P. and Nadeau, K.C., (2014), *Epigenetic regulation of asthma and allergic disease*, Allergy, Asthma & Clinical Immunology, **10**, 27.

Cantani, A., (1998), *Anaphylaxis from peanut oil in infant feedings and medications*, European Review for Medical and Pharmacological Sciences, **2**, 203–206.

Greer, F.R., Sicherer, S.H. and Burks, A., (2008), *Effects of Early Nutritional Interventions on the Development of Atopic Disease in Infants and Children: The Role of Maternal Dietary Restriction, Breastfeeding, Timing of Introduction of Complementary Foods, and Hydrolyzed Formulas*, Pediatrics, **121**(1), 183–192.

Hanson, L.A. et al., (1996), *Effects of breastfeeding on the baby and on its immune system*, Food Nutr. Bull., **17**, 384–9.

Whiteley, P. et al., (2010), *How Could a Gluten- and Casein-Free Diet Ameliorate Symptoms Associated with Autism Spectrum Conditions?*, Autism Insights, **2**, 39–53.

Greer, F.R., (2001), *Feeding the Premature Infant in the 20th Century*, J. Nutr., 131(2), 426S–430S.

Levin, M. et al., (2014), *Multiple independent IgE epitopes on the highly allergenic grass pollen allergen*, Clinical & Experimental Allergy, **44**, 1409–1419.

Kere, J., (2005), *Mapping and identifying genes for asthma and psoriasis*, Phil. Trans. R. Soc. B., **360**, 1551–1561.

Theije, C.G.M. de, et al., (2011), *Pathways underlying the gut-to-brain connection in autism spectrum disorders as future targets for disease management*, Eur. J. Pharmacol., **668**, Supplement 1, Pages S70–S80.
Castilho, S.D., Barros Filho A.A., (2010), *The history of infant nutrition*, J Pediatr *(Rio J)*, **86**(3):179–188.
Simopoulos, A.P., (2001), *The Mediterranean diets: What is so special about the diet of Greece? The scientific evidence*, J. Nutr., **131**(11 Suppl.): 3065S–73S.
Stevens, E.E., Patrick, T.E., Pickler, R., (2009), *A History of Infant Feeding*, The Journal of Perinatal Education, **18**(2), 32–39.
Arango, M.T., Perricone, C., Kivity, S. et al., (2017), *HLA-DRB1 the notorious gene in the mosaic of autoimmunity*, Immunol Res., **65**, 82–98, doi: 10.1007/s12026-016-8817-7.
Côté, C.D., Zadeh-Tahmasebi, M., Rasmussen, B.A. et al., (2014), *Hormonal Signaling in the Gut*, J. Biol. Chem., **289**, 11642–11649, doi: 10.1074/jbc.O114.556068.
Ballard, O., Morrow, A.L., (2013), *Human Milk Composition: Nutrients and Bioactive Factors*, Pediatr Clin North Am., **60**(1), 49–74, doi: 10.1016/j.pcl.2012.10.002.
Kewalramani, A., Bollinger, M.E., (2010), *The impact of food allergy on asthma*, Journal of Asthma and Allergy, **3**, 65–74.
Victora, C.G., (1996), *Infection and disease: The impact of early weaning*, Food and Nutrition Bulletin, **17**(4) (UNU, 163 pages).
Stoll, G., Bendszus, M., *Inflammation and Atherosclerosis Novel Insights into Plaque Formation and Destabilization*, (2006), Stroke, **37**(7), 1923–32.
Lue, L-F., Andrade, C., Sabbagh, M. et al., (2012), *Is There Inflammatory Synergy in Type II Diabetes Mellitus and Alzheimer's Disease?* International Journal of Alzheimer's

Disease, (Article ID 918680, 9 pages), doi: 10.1155/2012/918680.
Liu, C., Zhang, J., Shi, G.P., (2016), *Interaction between allergic asthma and atherosclerosis*, Transl. Res., **174**, 5–22.
Schluep Campo, I., and Beghin, J.C., (2005), *Dairy Food Consumption, Production, and Policy in Japan*, CARD Working Papers, Paper 418.
Inoue, M. et al., (2012), *Infant feeding practices and breastfeeding duration in Japan: A review*, International Breastfeeding Journal, **7**, 15, doi: 10.1186/1746-4358-7-15.
Leveque, L. and Khosrotehrani, K., (2011), *Can maternal microchimeric cells influence the fetal response toward self antigens?* Chimerism, **2**(3), 71–77, doi: 10.4161/chim.2.3.17589.
Wilding, J.P.H., (2002), *Neuropeptides and appetite control* ©Diabetes UK, Diabetic Medicine, **19**, 619–627.
Fox, E., Amaral, D., Van de Water, J., (2012), *Maternal and Fetal Anti-brain Antibodies in Development and Disease*, Dev. Neurobiol., **72**(10), 1327–1334. doi: 10.1002/dneu.22052.
Braunschweig D., Van de Water J., (2012), *Maternal autoantibodies in autism*, Arch. Neurol., **69**(6), 693–9, doi: 10.1001/archneurol.2011.2506.
Cook-Mills J.M., (2015), *Maternal Influences over Offspring Allergic Responses*, Curr. Allergy Asthma Rep., **15**(2): 501, doi: 10.1007/s11882-014-0501-1.
Hasselquist, D., Nilsson, J.A., (2009), *Maternal transfer of antibodies in vertebrates: trans-generational effects on offspring immunity*, Philos. Trans. R. Soc. Lond. B. Biol. Sci., **364** (1513), 51–60, doi: 10.1098/rstb.2008.0137.
Jeanty, C., Derderian, S.C., Mackenzie, T.C., (2014), *Maternal-fetal cellular trafficking: clinical implications and consequences*, Curr Opin Pediatr., **26**(3): 377–382, doi: 10.1097/MOP.0000000000000087.

Myhill, S., Booth, N.E., McLaren-Howard, J., (2013), *Targeting mitochondrial dysfunction in the treatment of Myalgic Encephalomyelitis/Chronic Fatigue Syndrome (ME/CFS) – a clinical audit*, Int. J. Clin. Exp. Med., **6**(1), 1–15.
Domínguez-López A., Miliar-García A., Segura-Kato Y.X., et al., (2005), *Mutations in MODY genes are not common cause of early-onset type 2 diabetes in Mexican families*, Journal of the Pancreas, (Online) **6**(2), 238–245.
Brown, E.S., and Chandler, P.A., (2001), *Mood and Cognitive Changes During Systemic Corticosteroid Therapy*, Prim Care Companion J. Clin. Psychiatry, **3**(1), 17–21.
Brugman, S., Perdijk, O., van Neerven, R.J., Savelkoul, H.F., (2015), *Mucosal Immune Development in Early Life: Setting the Stage*, Arch. Immunol. Ther. Exp. (Warsz). **63**(4), 251–68, doi: 10.1007/s00005-015-0329-y.
Medzhitov, R., (2007), *Recognition of microorganisms and activation of the immune response*, Nature, **449**, 819–826, doi: 10.1038/nature06246.
Caicedo, R.A., Li, N., Des Robert, C., et al., (2008), *Neonatal formula feeding leads to immunological alterations in an animal model of type 1 diabetes*, Pediatr Res. **63**(3), 303–7.
Maródi, L., (2006), *Neonatal innate immunity to infectious agents.* Infect Immun., **74**(4):1999–2006.
Palmer, A.C., (2011), *Nutritionally Mediated Programming of the Developing Immune System*, Adv. Nutr., **2**, 377–395.
Laugerette, F., Furet, J.P., Debard, C., et al., (2012), *Oil composition of high-fat diet affects metabolic inflammation differently in connection with endotoxin receptors in mice*, Am. J. Physiol. Endocrinol. Metab., **302**(3), E374–E386,. doi: 10.1152/ajpendo.00314.2011.
Slack, J.M.W., (1995), *Developmental biology of the pancreas*, Development, **121**, 1569–1580.

Rocca, W.A., Petersen, R.C., Knopman, D.S., et al., (2011), *Trends in the incidence and prevalence of Alzheimer's disease, dementia, and cognitive impairment in the United States*, Alzheimer's & Dement., **7**(1), 80–93, doi: 10.1016/j.jalz.2010.11.002.

Schmid-Ott, G., (2003),*Future Trends in Psychodermatological Psoriasis Research: Somatopsychic or Psychosomatic Focus?*, Dermatol. Psychosom., **4**, 129–130.

Alvarado, L.C., (2013),*Do evolutionary life-history trade-offs influence prostate cancer risk? A review of population variation in testosterone levels and prostate cancer disparities*, Evol. Applv., **6**(1), 117–133.

Ayala-Fontánez, N., Soler, D.C., McCormick, T.S., (2016), *Current knowledge on psoriasis and autoimmune diseases*, Psoriasis: Targets and Therapy, 2016, 6.

Hsing, A.W., Devesa, S.S., Jin F, Gao, Y.T., (1998), *Rising incidence of prostate cancer in Shanghai, China*, Cancer Epidemiol. Biomarkers Prev., **7**(1), 83–4.

Sunami, E., Kanazawa, H., Hashizume, H., et al., (2001), *Morphological characteristics of Schwann cells in the islets of Langerhans of the murine pancreas*, Arch. Histol. Cytol., **64**(2), 191–201.

Ciriaco, M., Ventrice, P., Russo, G., (2013), *Corticosteroid-related central nervous system side effects*, J Pharmacol Pharmacother, **4**(Suppl.): S94–S98, doi: 10.4103/0976-500X.120975.

Dahl, R., (2006), *Systemic side effects of inhaled corticosteroids in patients with asthma*, Respir. Med., **100**(8):1307–17.

Royer, C.M., Rudolph, K., Barrett, E.G., (2013), *The neonatal susceptibility window for inhalant allergen sensitization in the atopically predisposed canine asthma model*, Immunology, **138**(4), 361–9, doi: 10.1111/imm.12043.

Pauwels, E.K., (2011), *The protective effect of the Mediterranean diet: focus on cancer and cardiovascular risk*, Med. Princ. Pract., **20**(2), 103-11, doi: 10.1159/000321197.

Heard, E., Martienssen, A., (2014), *Transgenerational Epigenetic Inheritance: Myths and Mechanisms*, Cell, **157**(1), 95–109.

Fiocchi, A. (Chair), Brozek, J., Schünemann, H., et al. ,(2010), *World Allergy Organization (WAO) Diagnosis and Rationale for Action against Cow's Milk Allergy (DRACMA) Guidelines*, World Allergy Organ J. **3**(4): 57–161, doi:10.1097/WOX.0b013e3181defeb9.

Martelli, A., De Chiara, A., Corvo, M. et al., (2002), *Beef allergy in children with cow's milk allergy; cow's milk allergy in children with beef allergy*, Ann Allergy Asthma Immunol., **89**(6 Suppl. 1), 38–43.

zur **Hausen, H. and Villiers, E.-M. de**, (2015), *Dairy cattle serum and milk factors contributing to the risk of colon and breast cancers*, Int. J. Cancer., **137**, 959–967.